Advances in Information Security

Volume 86

Series Editor
Sushil Jajodia, George Mason University, Fairfax, VA, USA

The purpose of the *Advances in Information Security* book series is to establish the state of the art and set the course for future research in information security. The scope of this series includes not only all aspects of computer, network security, and cryptography, but related areas, such as fault tolerance and software assurance. The series serves as a central source of reference for information security research and developments. The series aims to publish thorough and cohesive overviews on specific topics in Information Security, as well as works that are larger in scope than survey articles and that will contain more detailed background information. The series also provides a single point of coverage of advanced and timely topics and a forum for topics that may not have reached a level of maturity to warrant a comprehensive textbook.

More information about this series at http://www.springer.com/series/5576

ElMouatez Billah Karbab • Mourad Debbabi
Abdelouahid Derhab • Djedjiga Mouheb

Android Malware Detection using Machine Learning

Data-Driven Fingerprinting and Threat
Intelligence

ElMouatez Billah Karbab
Security Researh Centre
Gina Cody School of Engineering and
Computer Science
Concordia University
Montreal, QC, Canada

Mourad Debbabi
Security Researh Centre
Gina Cody School of Engineering and
Computer Science
Concordia University
Montreal, QC, Canada

Abdelouahid Derhab
Center of Excellence in Information
Assurance Department
King Saud University
Riyadh, Saudi Arabia

Djedjiga Mouheb
Department of Computer Science
College of Sciences
University of Sharjah
Sharjah, United Arab Emirates

ISSN 1568-2633   ISSN 2512-2193   (electronic)
Advances in Information Security
ISBN 978-3-030-74666-7   ISBN 978-3-030-74664-3   (eBook)
https://doi.org/10.1007/978-3-030-74664-3

This Springer imprint is published by the registered company Springer Nature Switzerland AG
The registered company address is: Gewerbestrasse 11, 6330 Cham, Switzerland

Contents

List of Figures

List of Tables

Chapter 1
Introduction

Mobile apps have become essential in our life and work, as many of the services we use are provided to us through mobile apps. Moreover, Android OS has become the dominant platform not only for mobile phones and tablets but also for Internet of Things (IoT) devices [1–4]. In this context, Google has launched Android Things [5], an Android OS for IoT devices, where developers benefit from the mature Android stack to develop IoT apps for thin devices [4–7]. In contrast to personal computers (non-mobile systems), mobile devices are equipped with sophisticated sensors, from cameras and microphones to gyroscopes and GPS [8]. These sensors enable a whole new world of applications for end-users [8] and generate big amounts of data with highly sensitive information. The sophistication of mobile devices and the ubiquitousness of IoT objects help to build a smart world, but also unleash unprecedented potential for cyber-threats. Such threats could be committed by adversaries who might gain access to sensitive information and resources through Android malware apps. In this context, protecting Android devices from malicious apps is of paramount importance. This raises the need for security solutions to protect users from malicious apps, which exploit the sophistication and the sensitive content of smart devices.

1.1 Motivations

The volume of malware is growing tremendously [9], millions per month. In 2013, there was about 30 million malware in the whole year [10]. In February 2017, the number of new malware variants has reached 94.6 million [11]. This phenomenal growth is due to the ease of development of malicious apps, especially, repackaging existing malicious apps to develop new variants. There are some discrepancies in the estimation of the actual daily new malware [10, 11], but they agree to a large extent. In 2013, the estimation on new malware was about 82k per day [10].

© The Author(s), under exclusive license to Springer Nature Switzerland AG 2021
E. B. Karbab et al., *Android Malware Detection Using Machine Learning*, Advances in Information Security 86, https://doi.org/10.1007/978-3-030-74664-3_1

However, in 2017, experts [12] estimated a figure of 250k new malware a day. Many security solutions have been proposed to defend against Android malware. For example, the vetting system in app markets such as Google Play[1] plays a crucial role in detecting Android malware. However, Android malicious apps are able to deceive the vetting systems[2] in app markets[3] and infect millions of devices,[4] which emphasizes the importance of effective Android malware detection capability. The challenge relates to processing, analyzing, and fingerprinting new malware binaries to produce analytics in a limited time window. In this respect, the book aims at answering the following questions: (1) How can we efficiently fingerprint malware in a large binary corpus? (2) How can we effectively detect malware? (3) How can we group the detected samples into malware families?

There is a clear need for solutions that defend against malicious apps in mobile and IoT devices with specific requirements to overcome the limitations of existing Android malware detection systems. First, the Android malware detection system should ensure a high accuracy with minimum false alarms. Second, it should be able to operate at different deployment scales: (1) Server machines, (2) Personal computers, (3) Smartphones and tablets, and (4) IoT devices. Third, detecting that a given app is malicious may not be enough, as more information about the threat is required to prioritize the mitigation actions. Knowledge on the type of attack could be crucial to prevent the intended damage. Therefore, it is essential to have a solution that goes a step further and attributes the malware to a specific family, which defines the potential threat that infected system is exposed to. Finally, it is necessary to minimize manual human intervention as much as possible and make the detection dependent mainly on the app sample for automatic feature extraction and pattern recognition. As malicious apps are quickly getting stealthier, the security analyst should be able to catch up with this trend. In this respect, for every new malware family, a manual analysis of the samples is required to identify its pattern and features that distinguish it from benign apps.

1.2   Objectives

In this book, we focus on fingerprinting malware, i.e. finding distinctive patterns in malware in order to be distinguished from benign samples. The main goal of this book is the detection and the family attribution of Android malicious apps. The developed frameworks and techniques are applied to Android malware. However, the framework is general enough to be accommodated to address malware on other

---

[1]https://play.google.com.

[2]https://tinyurl.com/y57twlbs.

[3]https://tinyurl.com/yx8vqld2.

[4]https://tinyurl.com/y5rwybst.

platforms. Toward achieving the aforementioned goals, we dedicate special attention to the following criteria:

- **Scalable Malware Fingerprinting**: In response to the increasing number and magnitude of malware attacks, in this book, we focus on proposing scalable solutions and techniques for malware fingerprinting, while maintaining high malware detection performance.
- **Resilient Malware Fingerprinting**: Malware developers employ various obfuscation techniques to thwart detection attempts. Therefore, obfuscation resiliency is crucial in modern malware fingerprinting. In this book, we put the emphasis on resiliency to code transformation and common obfuscation techniques in the development of our malware fingerprinting techniques and systems.
- **Portable Malware Fingerprinting**: (1) Manual feature engineering of platform-dependent malware features is not scalable given the amount and the changing velocity of malware techniques. Therefore, portable feature engineering is an essential criterion in the development of fingerprint solutions, such as feature engineering in dynamic analysis. (2) The efficiency of the fingerprinting techniques, starting from the initial data processing to the detection, is a crucial factor that affects the deployment portability of the solution. (3) The diversity of targeted architectures, platforms, and execution environments of malware is a challenging problem.

1.3   Research Contributions

In this section, we present a summary of the book's detection solutions for Android malware.

- We propose APK-DNA [13], a fuzzy fingerprinting approach that captures both the structure and the semantics of the *APK* file using most Android *APK* features. We leverage APK-DNA to develop Cypider [14] framework, a set of techniques and tools aiming to perform a systematic detection and grouping of Android malware by building an efficient and a scalable similarity network of malicious apps. Our detection method is based on a novel concept, namely *malicious community*, in which we consider, for a given family, the instances that share common features. Under this concept, we assume that multiple similar Android apps with different authors are most likely to be malicious. Cypider leverages this assumption for the detection of variants of known malware families and zero-day malware. Cypider applies *community detection algorithms* on the similarity network, which extracts sub-graphs considered as suspicious and most likely malicious communities. We propose a novel fingerprinting technique, namely *community fingerprint*, based on a learning model for each malicious community.
- We elaborate, MalDy [15, 16], a portable malware detection and family threat attribution framework using supervised machine learning techniques. The key idea of MalDy portability is the modeling of the behavioral reports as a sequence

of words, along with advanced Natural Language Processing (NLP) and Machine Learning (ML) techniques for automatic engineering of relevant security features to detect and attribute malware without human in the loop. More precisely, we use the *bag-of-words* (BoW) model to capture the behavioral reports. Afterward, we build ML ensembles on top of BoW features. We evaluate MalDy on various datasets from different platforms (Android and Win32) and execution environments. The evaluation shows the effectiveness and the portability of MalDy across a spectrum of analyses and settings.

- We propose ToGather [17, 18], an automatic investigation framework that takes Android malware samples as input and produces insights about the underlying malicious cyber-infrastructures. ToGather leverages state-of-the-art graph analyses techniques to generate actionable, relevant, and granular intelligence to detect the threat effects induced by the malicious Internet activity of Android malware apps. The main contributions are: (1) We design and implement ToGather, a simple, yet practical framework for the generation of actionable, relevant, and granular intelligence on the malicious cyber-infrastructures used by Android malware. (2) We propose a correlation mechanism with multiple cyber-threat intelligence feeds, which enriches not only the resulting malicious cyber-infrastructure intelligence but also the labeling of the tracked malicious activities.

- We propose MalDozer [19–21], an automatic Android malware detection and family attribution framework that relies on method call sequence classification using deep learning techniques. Starting from a raw sequence of the app's API method calls, MalDozer automatically extracts and learns the malicious and the benign patterns from the actual samples in order to detect Android malware. MalDozer can serve as a ubiquitous malware detection system that can be deployed not only on servers but also on mobile and even IoT devices. This framework consists of the following: (1) MalDozer, a novel, effective, and efficient Android malware detection framework using the raw sequences of API method calls in conjunction with neural networks. In this context, we take a step beyond malware detection by attributing the detected Android malware to its family with high accuracy. (2) An automatic feature extraction technique during the training using *method embedding*, where the input is the raw sequence of API method calls, extracted from Android Dalvik assembly.

- We propose PetaDroid, a resilient and adaptive framework for Android malware detection and family clustering using advanced natural language processing and machine learning techniques. PetaDroid detects Android malware samples using an ensemble of Convolutional Neural Networks (CNNs) on top of our Inst2Vec features. Afterward, PetaDroid clusters the detected malware into groups of the same family utilizing sample digests generated using deep neural auto-encoder. PetaDroid is robust to common obfuscation techniques due to our fragment randomization technique during the training. PetaDroid leverages the confidence of detection decisions during deployment to collect extension dataset at each epoch. The extension dataset is used to automatically build new models without manual sample collection and also to empower time resiliency.

1.4 Book Organization

The remainder of the book is organized as follows: Chapter 2 provides the necessary background and state-of-the-art to Android malware detection. In Chap. 4, we propose an approximate fingerprinting approach for malware detection and apply it to Android malware. Afterward, we present a novel malware clustering technique, based on the aforementioned fingerprinting approach and graph partition techniques. Chapter 5 tackles portable malware fingerprinting from dynamic analysis reports using natural language processing and machine learning techniques. In Chap. 6, we propose a framework for Android malware cyber-infrastructure investigation. In Chap. 7, we propose a portable Android malware detection using deep learning techniques. In Chap. 8, we present an Android malware detection framework with high obfuscation resiliency and change overtime adaptation. Finally, Chap. 9 provides the conclusions, which discuss the relevance and importance of the addressed problems, followed by a summary of our contributions. In addition, it mentions the limitations of our research while presenting some avenues for future research on the topics studied.

References

1. Android Things on the Intel Edison board, https://tinyurl.com/gl9gglk. Accessed April 2016
2. Android Things operating system, https://developer.android.com/things. Accessed March 2020
3. RaspberryPI 3, https://www.raspberrypi.org/products/raspberry-pi-3-model-b/. Accessed Dec 2017
4. RaspberryPI 2, https://www.raspberrypi.org/products/raspberry-pi-2-model-b/. Accessed Jan 2017
5. Android Things, https://developer.android.com/things/. Accessed Sept 2016
6. Android Auto, https://www.android.com/auto/. Accessed April 2016
7. Android Wear Operating System, https://www.android.com/wear/. Accessed March 2016
8. F. Delmastro, V. Arnaboldi, M. Conti, People-centric computing and communications in smart cities. IEEE Commun. Mag. **54**(7), 122–128 (2016)
9. Malware Statistics and Trends Report, AV-TEST, https://www.av-test.org/en/statistics/malware. Accessed Jan 2020
10. Report: Average of 82,000 new malware threats per day in 2013. https://www.pcworld.com/article/2109210/report-average-of-82-000-new-malware-threats-per-day-in-2013.html. Accessed March 2014
11. Symantec Intelligence for February 2017, https://www.symantec.com/connect/blogs/latest-intelligence-february-2017. Accessed Oct 2017
12. New Malware Variants Near Record-Highs: Symantec, http://www.securityweek.com/new-malware-variants-near-record-highs-symantec. Accessed March 2017
13. E.B. Karbab, M. Debbabi, D. Mouheb, Fingerprinting Android packaging: generating DNAs for malware detection. Digit. Investig. **18**, S33–S45 (2016)
14. E.B. Karbab, M. Debbabi, A. Derhab, D. Mouheb, Cypider: building community-based cyber-defense infrastructure for android malware detection, in *Proceedings of the 32nd Annual Conference on Computer Security Applications, ACSAC 2016, Los Angeles, CA, USA*, 5–9 Dec 2016, pp. 348–362

15. E.B. Karbab, M. Debbabi, MalDy: portable, data-driven malware detection using natural language processing and machine learning techniques on behavioral analysis reports. Digit. Investig. **28**, S77–S87 (2019)
16. E.B. Karbab, M. Debbabi, Portable, data-driven malware detection using language processing and machine learning techniques on behavioral analysis reports. CoRR, abs/1812.10327 (2018)
17. E.B. Karbab, M. Debbabi, Automatic investigation framework for android malware cyber-infrastructures. CoRR, abs/1806.08893 (2018)
18. E.B. Karbab, M. Debbabi, ToGather: automatic investigation of android malware cyber-infrastructures, in *Proceedings of the 13th International Conference on Availability, Reliability and Security, ARES 2018, Hamburg, Germany*, 27–30 Aug 2018, pp. 20:1–20:10
19. S. Alrabaee, E.B. Karbab, L. Wang, M. Debbabi, BinEye: towards efficient binary authorship characterization using deep learning, in *Proceedings, Part II Computer Security - ESORICS 2019 - 24th European Symposium on Research in Computer Security, Luxembourg*, 23–27 Sept 2019, pp. 47–67
20. E.B. Karbab, M. Debbabi, A. Derhab, D. Mouheb, Android malware detection using deep learning on API method sequences. CoRR, abs/1712.08996 (2017)
21. E.B. Karbab, M. Debbabi, A. Derhab, D. Mouheb, MalDozer: automatic framework for android malware detection using deep learning. Digit. Investig. **24**, S48–S59 (2018)

Chapter 2
Background and Related Work

In this chapter, we review and compare the state-of-the-art proposals on Android malware analysis and detection according to a novel taxonomy. Due to the large number of published contributions, we focus our review on the most prominent articles in terms of novelty and contributions, with an emphasis on those published in top-tier security journals and conferences. The proposed taxonomy is based on the generality of Android malware threats. It classifies the existing systems into: (1) *general malware detection*, which aims to detect malware without taking into account a particular type of attack, and (2) *attack-based malware detection*, which aims at detecting specific attacks such as privilege escalation attacks, data leakage attacks, etc. Furthermore, each threat category is classified according to the system deployment of the detection approach, i.e., the physical environment into which the system is intended to run. Furthermore, we consider three main deployment architectures: *workstation-based*, *mobile-based*, and *hybrid* architectures. The proposed two-level taxonomy allows carrying out an objective and appropriate analysis by comparing only systems that are addressing the same threat category, and having the same deployment architecture as they share the same goals and have similar issues to solve.

2.1 Background

In this section, we introduce the essential background knowledge of Android OS. We also discuss briefly Android security and its implication on malware detection.

© The Author(s), under exclusive license to Springer Nature Switzerland AG 2021
E. B. Karbab et al., *Android Malware Detection Using Machine Learning*, Advances in Information Security 86, https://doi.org/10.1007/978-3-030-74664-3_2

2.1.1 Android OS Overview

Android is a mobile operating system maintained by Google and supported by the Open Handset Alliance (OHA) [1]. Android is embraced by the Original Equipment Manufacturers (OEMs), chip-makers, carriers, and application developers. Android apps are written in Java. However, the native code and shared libraries are developed in C/C++ [2]. The current Android architecture [3] consists of the Linux kernel, which is designed for an embedded environment with limited resources. On top of the Linux kernel, there is the Hardware Abstraction Layer (HAL), which provides standard interfaces that expose device hardware capabilities to the higher-level Java API framework, by allowing programmers to create software hooks between the Android platform stack and the hardware. Also, there is Android Runtime (ART), which is an application runtime environment used by the Android OS and which replaced Dalvik virtual machine starting from Android 5.0. ART translates the apps' bytecode into native instructions that are later executed by the device's runtime environment. ART introduces the Ahead-Of-Time (AOT) compilation feature, which allows compiling entire applications into native machine code upon their installation. The native libraries developed in C/C++ support high-performance third-party reusable shared libraries. The Java API framework provides APIs for the building blocks the user needs to create Android apps.

2.1.1.1 Android Apk Format

Android Application Package (*Apk*) is the file format adopted by Android for apps distribution and installation. It comes as a *ZIP* archive file, which contains all the components needed to run the app. By analogy, *Apk* files are similar to Windows *EXE* installation files or Linux *RPM*/*DEB* files. The *Apk* package is organized into different directories (namely, **lib**, **res**, **assets**), and files (namely, **AndroidManifest.xml** and **classes.dex**). More precisely, the **AndroidManifest.xml** file contains the app meta-data, e.g., name, version, required permissions, and used libraries. The **classes.dex** file contains the compiled Java classes. The **lib** directory stores C/C++ native libraries [2]. The resources directory (**res**) contains the non-source code files, such as video, image, and audio files, which are packaged during compilation.

2.1.1.2 Android Markets

Android app market is an Internet site for developers to publish their apps. Google Play is the official app market for Android. Before an app is published in this market, it needs to be verified by the Bouncer vetting system [4] to check newly submitted apps against malware. This involves scanning an app for known malicious code and performing dynamic analysis for a limited period to detect hidden malicious behaviors. The vetting system can be evaded by apps that avoid triggering malicious behavior during the analysis time. There are also other third-party markets such as AppChina [5] and Mumayi [6], where developers can upload their apps. However, they provide fewer restrictions to publish apps. Unlike Google Play store, they do not vet the submitted apps but rather rely on users' feedback, which helps attackers to publish repackaged apps and malware easily (minimum vetting).

2.1.2 Android Security

Android OS employs two security mechanisms: permissions and sandboxing. In Android, apps can access resources such as telephony, network, and SMS functions using APIs. Android APIs are protected using a security mechanism based on permissions. Each application must define the permissions it requests in its AndroidManifest.xml file. A user needs to grant the required permissions to install the app. Otherwise, the application cannot be installed. The Android kernel provides a sandboxing feature, which isolates apps from one another. In Android, each application is assigned a unique User ID (UID) and is run as a separate process. The file system access policy does not allow one user (resp., application) to access or modify another user's (application's) files.

2.1.2.1 Android Security Threats

Attackers can exploit many weaknesses and vulnerabilities in the Android ecosystem to compromise and infect Android devices with malware. These weaknesses are summarized as follows: (1) Most of the App markets deploy no or limited vetting system to check whether the submitted apps are malicious or not. (2) The official Android market might contain malicious apps. Some apps [7–9] in Google's Play Store have been identified as malicious. Thus, there is a risk that users download malware. (3) There are non-market sources to obtain Android apps such as SD cards, which open more attack entry points for malware. (4) Most of the users ignore or have little understanding of the Android permission policy. This is stated in a survey [10], which showed that only 17% of users look at the permissions when installing applications. (5) It has also been noticed that developers request more permissions than they need [11]. As legitimate and malicious applications can request permissions, it is often difficult for users to determine, during the installation time whether the requested permissions are harmful or not. (6) It is relatively easy for an adversary to reverse engineer a legitimate Android app, insert malicious code, and repackage the Apk file again.

2.1.2.2 Design Challenges of Malware Detection Systems

The design of Android malware detection system faces many challenges: (1) Ensuring simultaneously high-accuracy detection and efficiency in terms of time and resource use (CPU, RAM, and battery in case of mobile device deployment), is difficult to achieve, especially in the case of deploying the detection system on resource-constrained devices. (2) Malware developers employ techniques to evade detection, such as code obfuscation or dynamically loading a binary code from a remote server. (3) The vetting system executes apps for a limited time in a controlled environment, e.g., sandboxing or emulation, to check their maliciousness. However, some malware only reveal their malicious behaviors after a period of time to escape the analysis of maliciousness detection systems during the early execution period. Also, some malware try to check the execution environment for signs of a sandboxing system [12, 13]. The goal is to prevent the execution of the malicious payload under a sandboxing environment.

2.2 Android Malware Detection Overview

In this section, we review the existing contributions on Android malware analysis. Android malware detection methods focus on identifying whether the analyzed app is benign or malicious. Malware detection proposals can be categorized into static [14, 15], dynamic [16, 17], and hybrid [18, 19] analysis-based.

Static Analysis Approaches Static analysis techniques perform fast code disassembling and decompilation without the need to execute the binary. The static methods [14, 15, 20–32] depend on static features, which are extracted from the Apk file such as requested permissions, APIs, bytecodes, opcodes. Static analysis techniques are fast and cover all the execution paths of the analyzed app. However, this approach is undermined by the use of various code transformation techniques [33]. We may divide static analysis based techniques into the following categories: (1) **Signature-based analysis:** This analysis deals with the extracted syntactic pattern features. The authors in [15] create a unique signature matching a particular malware. However, such signature cannot handle new variants of existing known malware. Moreover, the signature database should be updated to handle new variants. AndroSimilar [34] has been proposed to detect zero-day variants of known malware. It is an automated statistical feature signature-based method for malware detection. (2) **Resource-based analysis:** The manifest file contains important metadata about the components, i.e., activities, services, receivers, etc. and the required permissions. There are some methods that have been proposed to extract such information and subject it to analysis [15, 20, 35, 36]. (3) **Permission-based analysis:** This approach is based on discovering unnecessary permission requests that might lead to malicious activity [37, 38]. In [39], the authors proposed a certification tool that defines a set of rules to detect malware by identifying combinations of requested permissions. (4) **Semantic-based analysis:** There are existing approaches that analyze Dalvik bytecode that is semantically rich, containing type information such as classes, methods, and instructions. Additionally, such information can be used to analyze control and data flow graphs that reveal privacy leakage and telephony services misuse [40, 41].

Dynamic Analysis Approaches The dynamic methods [16, 17, 42–46] use features that are derived from the app's execution. They are more resilient to code obfuscations than static analysis methods. However, such methods [45, 47–52] incur additional cost in terms of processing and memory to run the app. Also, anti-emulation techniques such as sandbox detection and delaying malware execution can evade dynamic analysis methods. Dynamic techniques are divided into the following two categories:

1. **Resources usage based:** Some malicious apps may cause Denial of Service (DoS) attacks by over-utilizing constrained hardware resources. A range of parameters such as CPU usage, memory utilization statistics, network traffic pattern, battery usage, and system calls for benign and malware apps are gathered from the Android subsystem. Then, automatic analysis techniques along with machine learning techniques are employed [50–52].
2. **Malicious behavior based:** This is related to abnormal behaviors such as sensitive data leakage and sending SMS/emails [45, 47–52].

Hybrid Analysis Approaches The hybrid methods [18, 19, 40, 53–59] use both static and dynamic features.

Malware Family Attribution In the previous categories, the proposed systems focus mainly on the detection task in which we segregate malware and benign apps. In this category, the proposed systems focus on the malware family attribution task, in addition to the detection task, as a goal for the analysis. Malware family attribution aims to attribute the malware to its actual family. To secure Android systems, some methods [60–62] focus on detecting variants of known families. Other proposals [63–76] adopt the unsupervised learning approach to find families of similar apps. These proposals assume that two or multiple apps that share similar code are likely to belong to the same malware family. Thus, they check if the apps are using the similar malicious code (i.e., detection of malware families), or they check for reused code of the same original app (i.e., code reuse detection).

2.3 Taxonomy of Android Malware Detection Systems

In this section, we present our taxonomy for Android malware detection systems. As shown in Fig. 2.1, the taxonomy considers three aspects for the classification and the comparative study: (1) *Targeted threat* (Fig. 2.2), (2) *Android stack layer* (Fig. 2.3), and (3) *System deployment* (Fig. 2.4) aspects are for classification. The first three aspects, i.e., targeted threat, system deployment, and the implementation layer, are

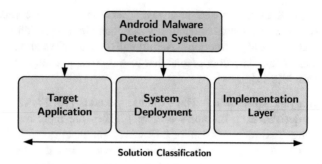

Fig. 2.1 Classification aspects

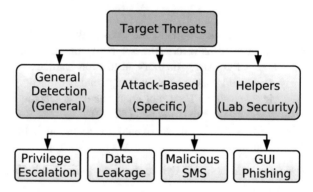

Fig. 2.2 Classification of target threats

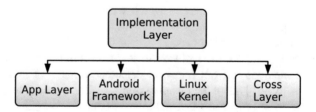

Fig. 2.3 Classification of the implementation layer

used to cluster related works that address the same threats in the same performance
objectives (i.e., they operate under the same class of deployment settings); finally
it defines on which Android OS stack layer, the system is implemented (i.e., *app*,
framework, and *Linux kernel* layer as shown in Fig. 2.3). Afterward, we conduct a
comparative study on the resulting groups for the criteria of the last two aspects, i.e.,
Feature Selection and Detection strategy.

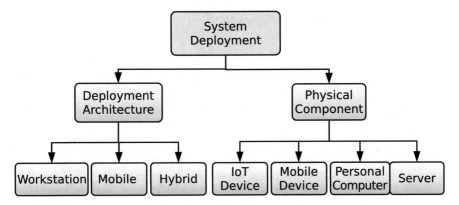

Fig. 2.4 System deployment classification

2.3.1   Malware Threats

Android Malware detection systems are designed to protect against different kinds of malicious activities. Based on the targeted threats, we classify them, as shown in Fig. 2.2, into: *Attack-based detection systems*, *generic detection systems*, and *helpers systems*. By (1) *attack-based detection systems* (also attack-dependent systems), we mean the systems that aim to address specific attacks. For instance, *Graphical User Interface (GUI) phishing* in which the adversary might develop a malicious app whose graphical user interface (GUI) is visually similar to a legitimate app, to deceive the user. In another example, two or more apps can collaborate to launch a *collusion attack* leveraging Android Inter-Components Communication (ICC). Although each app has a set of non-critical permissions, the apps can collude to generate a joint set of permissions that enables them to perform unauthorized malicious activities.

By *generic detection systems*, we mean systems that aim to detect malware without considering specific attacks. Instead, they detect malware based on prior-knowledge of specific malware patterns or a learned model. Alternatively, other detection systems focus on apps that contain similar code or exhibit similar behaviors. On the other hand, *Helpers systems* provide assistance to Android malware detection solutions. First, they are mainly used in lab environments due to their need for heavy manual work. Second, the provided tool or results of such assets will be leveraged to improve and enhance the actual malware detection solutions such as code obfuscation tools. Finally, these assets may not have been directly related to malware detection, but they could help sharpen the malware detection rate.

2.3.2 Detection System Deployment

Detection system deployment refers to the adopted architecture in terms of physical components. An Android malware detection system can be deployed on different types of architectures: (1) *workstation*, (2) *Mobile*, and (3) *hybrid*, as shown in Fig. 2.4. Each type of architecture has its design objectives and operates under specific constraints and deployment settings, as described below:

Workstation-Based Architecture This type of architecture is centralized because all the detection modules are deployed on a high-power server or high-end desktop machine, and it can be used for two types of application scenarios:

- First, *App market analysis*, in which the detection system, as depicted in Fig. 2.5a, has to check a newly submitted app before publishing it on the market. It represents the first line of defense in the Android ecosystem. For this reason, high detection accuracy is required. The second requirement is the scalability of the detection system with respect to detection time to the high arrival rate of apps, and hence the detection should be performed in real-time, i.e., online detection mode.
- Second, *Security lab analysis*, the security lab analyst can be any user working in the industry, a researcher from academia, or independent analyst, who uses machines of middle-resource capability regarding CPU and memory such as desktop computers and laptops, as depicted in Fig. 2.5b. We need this separate category because the malware analysis may involve manual investigation by the security practitioner; in contrast, app market analysis category has fully automated detection process. Thus, it is important to stress that this category leverages experts knowledge and may involve manual analysis.

Mobile-Based Architecture Any user with a mobile device can perform this analysis. The detection operations (provided by preprocessing, feature extraction, and the detection components) are carried out on the mobile device, as depicted in Fig. 2.6a. In addition to ensuring high detection accuracy, the detection system should cope with the limited capabilities of mobile devices that are characterized by constrained resources such as CPU, memory, and battery capacity.

Hybrid Architecture If the malware detection system operations are split between the mobile device and a workstation, the architecture is called hybrid (see Fig. 2.6b). In this case, the system should incur low communication cost in addition to consuming limited resources in terms of CPU, memory, and battery.

As each type of deployment system has different priority design objectives and offers different resource capacities from one another, we compare the detection systems that are categorized under the same deployment type. The differences between deployment systems regarding the importance of design objectives are summarized in Table 2.1.

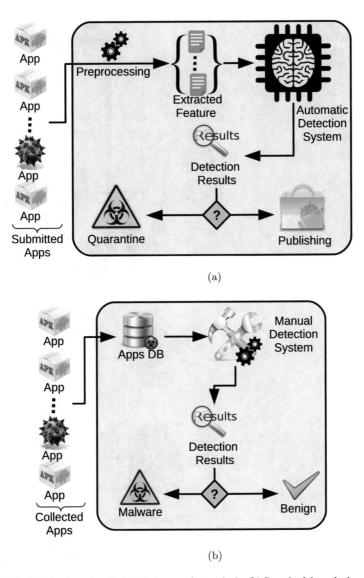

Fig. 2.5 Workstation-based analysis. (**a**) App market analysis. (**b**) Security lab analysis

2.4   Performance Criteria for Malware Detection

In this section, we present a generic framework as a template for Android malware detection systems together with the performance criteria. This template helps to make an abstraction of the components of a typical Android malware detection system and its characteristics and criteria based on its position in the proposed taxonomy (previous section). The framework, as shown in Fig. 2.7, consists of two

Fig. 2.6 Hybrid and
mobile-based analysis. (**a**)
Mobile-based detection. (**b**)
Hybrid detection

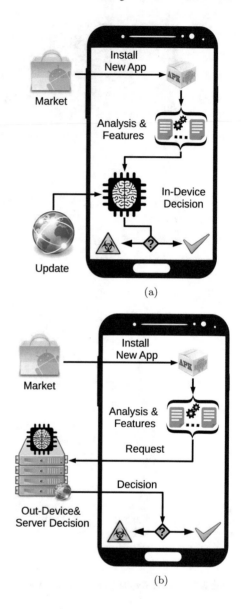

main processes: *Process to generate detection patterns*, which builds the detection
pattern (or model), and *Detection and response process*, which analyzes a target Apk
based on the developed detection model. In general, a dataset of Apks is the input
for the Detection pattern generation process. The Apk is first processed through
disassembling or decompilation to generate the raw features as they exist in the Apk
file. In some cases, the raw features are processed to generate high-level features,
which are, in turn, fed to training modules to produce the detection model. Given a

Table 2.1 Design objectives of deployment systems

Architecture deployment	Workstation-based		Mobile-based	Hybrid
	App market analysis	Security lab analysis		
High accuracy	Very important	Important	Very important	Very important
Detection time	Very important	Important	Important	Important
CPU	Less important	Important	Very important	Very important
RAM	Less important	Important	Very important	Very important
Battery	Not important	Not important	Very important	Very important
Communication	Not important	Not important	Not important	Very important
Usability	Not important	Not important	Important	Important

target Apk, the detection module will determine whether it is malicious or benign. The detection result can take one of the following two forms: label (i.e., malicious or not) and score (i.e., risk score). Also, a detection response might follow after obtaining the detection result.

2.4.1 Feature Selection

As shown in Fig. 2.8, features can be (1) Static, extracted from the Apk file, such as permissions, API used, opcode, etc. (2) Dynamic, which is extracted from the running the app, such as system calls, invoked APIs, network traffic, etc. (3) Hybrid, which combines both static and dynamic features. The features can also be classified with respect to other aspects as follows:

Code Transformation and Obfuscation Resiliency An obfuscation technique aims to evade detection by instrumenting the features used by the detection model. The feature is said to be highly resilient to an obfuscation technique if the malware detection process using such a feature is less affected by such technique, or the malware needs to change its functional logic to evade detection.

Adaptation to OS and Malware Evolution The release of a new Android OS version implies a new set of API frameworks. Detection systems that consider APIs as features need to manually redefine the set of API features before applying the detection process, which makes the adaptation of the detection system difficult.

Features Preprocessing Complexity In static analysis, before extracting some raw features (APIs used, opcode), it is required to disassemble the dex file, which takes a longer time than removing permissions, which only requires accessing the Manifest file. Although permissions are quickly extracted, the detection methods mainly based on such features are less resilient to obfuscations compared to those employing features that are derived from the dex file.

Detection Required Runtime This measures the time interval between processing the Apk and making a decision (malware or not) about a given Android app.

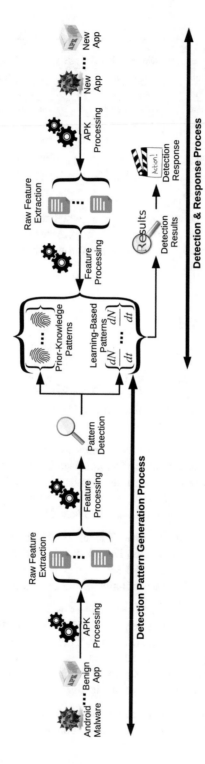

Fig. 2.7 Template framework for android malware detection systems

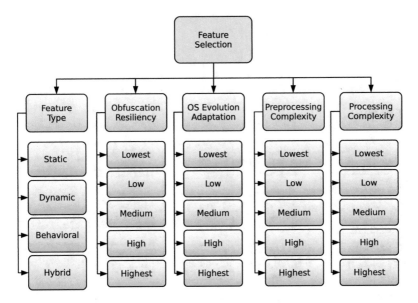

Fig. 2.8 Feature selection criteria

2.4.2 Detection Strategy

As shown in Fig. 2.9, the detection can be either offline or online. Upon malware detection, the *detection response* can be either: passive notification sent to the user, active reaction, which blocks the malware, or no response in the case of App market analysis and security lab analysis. The *detection scope* defines the goals of the Android malware fingerprinting: only malware detection, family attribution, or go a step further such as the detection of threat network, composed of IP addresses and domains names, related to Android malware samples. The *detection approach* indicates how the detection model is produced, which can be through a learning procedure or a prior-knowledge. The prior-knowledge models are specification-based models that are manually constructed by a human expert based on rules that try to determine the legitimate or the malicious behavior of the app. The main advantage of the prior-knowledge techniques is short detection as they only check if the predefined rules are violated or not. The main drawback is that building knowledge requires high-level human expertise and is often time-consuming and a difficult task. As for the learning models, they are automatic and can adapt to changes when new information is acquired.

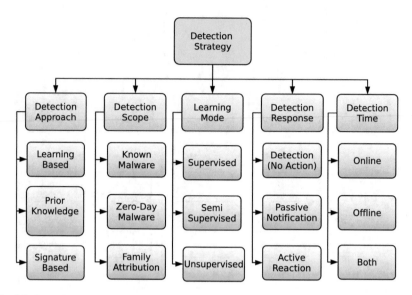

Fig. 2.9 Detection strategy comparison criteria

2.5 General Malware Threat Detection

In this section, we discuss Android malware detection proposals in the literature that target general malware detection without focusing on: *type of malware*, and *its attack techniques*. Moreover, this category leverages general features to fingerprint the malware. We classify the proposals in this category into *workstation-based*, *mobile-based*, and *hybrid-based* solutions. This section comes after we define our classification taxonomy and comparison criteria to describe and compare between the solutions in the light of our taxonomy and criteria.

2.5.1 Workstation-Based Solutions

This section presents the *workstation-based* solutions, which require a lot of resources in terms of processing and memory, so its deployment must be on relatively powerful machines.

MAMADROID [77, 78] is a framework for Android malware detection using static behavioral features of Android apps. MAMADROID leverages static analysis to extract learning features from a reverse engineered Android app, and Android APIs call sequences (a sequence of API is a possible behavior of the Android app). Markov chains are used to model and represent abstract forms of API sequences; the abstracted APIs consists of (1) Package such as java.lang and (2) Family such as java. In this model, a node is an Android API, and an edge is the probability of having a transition from one API to another. The authors consider these probabilities as features for the MAMADROID machine learning system. Therefore, Markov chains play the role of features extractor, and it is the cornerstone of the proposed detection system. MAMADROID can only be deployed in large-scale and high-end machines due to the required resources. It is relatively slow due to intense preprocessing; the reported runtime might reach 13–18 min depending on the size of the binary.

Stormdroid [79] is a malware detection system that considers malware as a stream of apps as in the case of App markets. The authors are motivated by the fact that most existing systems rely mainly on permissions and sensitive Android APIs. Therefore, they propose Stormdroid as a machine learning system for Android malware detection. Stormdroid leverages both static and dynamic analysis to achieve a high detection rate with a low false-positive rate. The evaluation of Stormdroid includes multiple types of machine learning classifiers; K-Nearest Neighbors is the most accurate one. However, K-Nearest Neighbors is not sufficiently scalable because the detection runtime grows linearly with the size of the training dataset; but it is constant for classification algorithms.

The main goal of OpSeq [60] is to measure the similarity of unknown apps relative to known malware. OpSeq defines similarity as a function of normalized opcode sequences found in sensitive functional modules as well as app permission requests. OpSeq extracts the components from a known sample and creates corresponding signatures, which are used to search for similar components in target applications. OpSeq improves the use of opcode-sequence similarity by focusing only on the components that make suspicious API calls. The list of requested permissions is used as a second parameter to improve detection accuracy. OpSeq is tested on a dataset of 1192 known malware samples belonging to 25 different families, 359 benign apps, and 207 new obfuscated malware variants. The results show that OpSeq can correctly detect known malware with an F1-Score of 98%. However, OpSeq generates fingerprints from only assembly opcodes N-grams from components that make suspicious API calls: (1) Relying on opcodes only ignores other information in the assembly code. (2) The system might not be resilient to the rapid change in suspicious API calls (how can find out new suspicious APIs in new OS versions).

2.5.2   Mobile-Based Solutions

This section presents *mobile-based* solutions. The following research initiatives target mobile deployment. In the mobile-based solutions, the design of the detection process (from the preprocessing, and feature extraction, to the final decision) is optimized to fit in resources-constrained devices with small memory capacity and computation power.

Drebin [14] is a supervised machine learning system for Android malware detection. Static analysis is used to extract a variety of code and app features from Dalvik and Manifest file, respectively. As for the code features, the authors rely mainly on the used permissions and Android APIs (or lightweight features), which have been filtered to consider only suspicious APIs, such as dynamic load or cryptographic API. App features, extracted from Manifest file, are mainly the requested permissions, the app components and intents, and the requested hardware components. Drebin only considers lightweight features so that the preprocessing would be very fast. Lightweight features extraction is essential for Drebin system efficiency to run on mobile devices (also high-end servers). Drebin leverages bit-vector to represent the extracted features in order to train a Support Vector Machine (SVM) model offline. Drebin [14] shows high detection performance with low false positive. In addition to detection results, Drebin provides explanatory details to the end-user in terms of suspicious features scores that affect the detection decision. However, the usage of lightweight features impacts the resiliency of the system against common code obfuscation techniques. Also, Drebin relies on manual feature engineering that produces a static list of features, which is less resilient to OS change overtime (new features due to the new APIs, permissions, etc.).

DroidBarrier [80] is a runtime process authentication model for Android. DroidBarrier provides legitimate apps (that are considered as benign at the launch of the system) with security credentials, which are used for authentication when the associated processes are created. Processes that do not have credentials fail to authenticate, and their corresponding applications are considered malicious. The main disadvantage of this approach is the need to determine which requests are initially legitimate.

A machine learning approach for a malware detection system that runs on Android devices is proposed in [81]. The authors model the detection system as a machine learning model, specifically as an anomaly detection system (only train the detection model on benign apps). To this end, they leverage a large number of available benign apps to train the anomaly detection model. Therefore, benign apps are considered as part of the normal area while everything else is considered as anomalies or malicious apps. The authors extract from the Android apps features such as permissions and Control Flow Graphs (CFG). A one-class support vector machine model, i.e., the anomaly detection model, is trained on the previous features offline. Afterward, the model is deployed on Android devices.

2.5.3 Hybrid Solutions

Hybrid solutions have two parts that are deployed, respectively, in mobile end-devices and the end-servers. Both ends work collaboratively to achieve detection of Android malware.

Crowdroid [48] is a detection system, which is composed of a lightweight app, that is installed on a mobile device and interacts with a centralized server to make detection decisions. The system, at the mobile device, monitors the Linux Kernel system calls of each app and sends the gathered information to a centralized server. The latter collects these calls from different users and applies a clustering algorithm to distinguish between benign and malicious apps. Although the reported detection performance of Crowdroid is perfect in some cases, the employed dataset is very small to the generalize the performance results, as shown in Table 2.3. Also, the usage of anomaly detection tends to generate a considerable amount of false positives compared with supervised classification.

DRACO, a hybrid Android malware detection system in [18], is proposed. In [18], the authors propose DRACO, which is a hybrid Android malware detection system. The system uses both dynamic and static analyses on two different ends: end-user smart devices and high-end servers, respectively. DRACO is hybrid because it is deployed on Android devices and servers simultaneously. In Android devices, DRACO collects dynamic analysis features such as the CPU, memory usage, and file system accesses. On the server-side, DRACO conducts a static analysis to extract features such as Android API calls. Both sets of features are used to train a support vector machine model as the core of the DRACO system. Using end-user smart devices as a dynamic analysis system raises the issue of privacy and security of DRACO itself, whether at the client or the server-side. Also, the used features are straightforward and are extracted completely in nowadays devices; therefore, this opens the question of the need for the servers.

Monet [82] is a runtime fingerprinting system for Android malware detection. Monet's generated fingerprints capture the behavior of the installed apps in the Android devices overtime. The authors design and implement Monet, a system that is deployed on both Android devices (client) and the high-end servers (server). Monet client is implemented in the Android middle-ware framework and Linux kernel to capture the dynamic behaviors of the apps. Afterward, the system summarizes and represents captured behaviors in the form of graphs or runtime fingerprints. These are later sent to the Monet server to match against known malware for detection purposes. Monet's client needs changes in both the Android framework in the Linux kernel to properly function, which makes the deployment very hard on a large scale and for simple end-users.

2.5.4 Discussions

Table 2.2 depicts a comparison between *generic* solutions relying on *feature selection* and *detection strategy* criteria. All the *workstation* solutions [60, 77, 79] employ static analysis to extract malicious and benign features because it is fast and allows to detect most malicious apps. These proposals are resilient to common code transformation techniques, depending on the strength of the chosen features [77]. However, these proposals are not immune to more advanced obfuscation techniques such as encryption, especially if the solution uses relatively simple static features[60]. For this reason, Stormdroid [79] leverages dynamic analysis to enhance the detection rate against obfuscated malicious apps. Stormdroid [79] and MAMADROID [77, 78] apply supervised machine learning techniques; in contrast,

Table 2.2 Qualitative comparison of general malware attack solutions

Solution	Feature selection			Detection strategy				Response
	Type	Resiliency	Adapt	Approach	Scope	Mode		
Stormdroid [79]	Hybrid	Highest	Medium	Learning	Known, zero-day	Supervised		–
Mamadroid [77]	Static	High	High	Learning	Known, zero-day	Supervised		–
OpSeq [60]	Static	Medium	Low	Signature	Known, attribution	–		–
Drebin [14]	Static	Medium	Low	Learning	Known, zero-day	Supervised		Passive
Droidbarrier [80]	Behavioral	Highest	High	Knowledge	–	–		Active
ML [81]	Static	Medium	Medium	Learning	Known	Supervised		Passive
Crowdroid [48]	Behavioral	Highest	High	Learning	Known, zero-day	Unsupervised		Passive
DRACO [18]	Hybrid	Highest	High	Learning	Known, zero-day	Supervised		Passive
Monet [82]	Hybrid	Highest	Highest	Signature	Known	–		Passive

Opsec [60] relies on fuzzy signatures to identify malicious apps. In the case of *mobile-based* solution, the authors of Drebin [14, 81] efficiently extract simple static analysis features (such as permissions) to minimize the needs in terms of resources; also, they apply supervised machine learning, classification [14] and anomaly detection [81]. Furthermore, [14] and [81] provide a passive notification to the end-user in contrast to Droidbarrier [80], which prevents malicious actions. To do so, Droidbarrier [80] leverages runtime behaviors in the mobile device to make a detection decision (authentification mechanism). *Hybrid* solutions [18, 48, 82] rely on behavioral features collected from mobile devices to provide passive alarms for the end-users.

Table 2.3 depicts quantitative comparison between *general* solutions. *Apk pre-processing complexity* varies between solutions. The highest complexity (Stormdroid [79]) is related to the dynamic analysis of an Apk file in a sandbox environment; also, this might require to repackage the Apk file [18] to include the monitoring APIs. A lower complexity [14] is achieved with decompressing Apk file. The lowest Apk processing complexity solution [80] could not use the Apk file because it monitors the running apps that are already installed in the mobile device. *Feature extraction complexity* could be less costly even though the solution uses dynamic analysis because the process of extracting is applied to simple logs files, in addition to the static features, such as Stormdroid [79], which requires a relatively small amount of time for the feature extraction. In contrast, MAMADROID [77] uses static analysis, yet its feature extraction is very complex and time-consuming

Table 2.3 Quantitative comparison of general malware attack solutions

Solution	Apk prepro-cessing complexity	Feature extraction complexity	Model generation computation	Detection rate	Detection time	Dataset
Stormdroid [79]	Highest	Medium	Lowest	Acc = 93.8%	201 s	Ben = 4.4k, Mal = 3.7k
Mamadroid [77]	High	Highest	Low	F1 = 99%	13–18 min	Ben = 8.5K, Mal = 35.5K
OpSeq [60]	Medium	Medium	Lowest	F1 = 98%	11.6 s	Ben = 0.4K, Mal = 1.2K
Drebin [14]	Low	Low	Low	Acc = 94%	10 s	Ben = 124K, Mal = 5.6K
Droidbarrier [80]	Lowest	Lowest	Lowest	–	–	Ben = 0K, Mal = 1.3K
ML [81]	Medium	High	Low	F1 = 85%	–	Ben = 2.1K, Mal = 0.1K
Crowdroid [48]	Low	Low	High	Acc = 85–100%	–	Ben = 0.05K, Mal = 0.01K
DRACO [18]	Highest	Low	Low	Acc = 98.4%	96 s	Ben = 18K, Mal = 10k
Monet [82]	High	High	Lowest	Acc = 99%	60 s	Ben = 0.5K, Mal = 3.8K

Table 2.4 Classifications of general malware attack solutions

Solution	Layer	Architecture	Physical components	Target application
Stormdroid [79]	App layer	Workstation	Server	Malware detection
Mamadroid [77]	App layer	Workstation	Server	Malware detection
OpSeq [60]	App layer	Workstation	PC	Malware detection
Drebin [14]	App layer	Mobile	Mobile/IoT devices	Malware detection
Droidbarrier [80]	Linux kernel	Mobile	Mobile/IoT devices	Malware detection
ML [81]	App layer	Mobile	Mobile devices	Malware detection
Crowdroid [48]	Framework, app layers	Hybrid	Mobile/IoT devices, server	Malware detection
DRACO [18]	App layer	Hybrid	Mobile/IoT devices, server	Malware detection
Monet [82]	App, framework layers	Hybrid	Mobile devices, server	Malware detection

(13–18 minutes per app), which is very high. The *model generation computation* could be very light in case of signature-based solutions [60, 82] or classification-based solutions [79] using *k-nearest neighbor* technique; because the solution needs only to fetch the most similar app in the signature database or the training set. Most solutions show very high detection rates. However, their evaluation dataset size variability drastically affects the generalization of the performance results. For instance, MAMADROID has 88k apps and Drebin has 130k apps in their dataset, while Crowdroid [48] has 0.06k apps. The detection rate, time, and dataset size in Table 2.5 are collected from the original publications of the solution.

As shown in Table 2.4, depending on the deployment architectures of the **general solutions**, we could classify them into *workstation*, *mobile*, and *hybrid* based architecture. *Workstation* solutions such as [60, 79] target Android app layer for malware detection. In contrast, Droidbarrier [80] (*mobile-based*), is implemented in the Linux Kernel layer. Furthermore, Monet [82] (*hybrid* solutions) is implemented across two layers, Android framework and app layers.

2.6 Specific Malware Threat Detection

In this section, we present the *attack-based* detection solutions. These solutions still consider the detection of malicious apps but they target a specific malicious behavior in the malicious app such as *sensitive data leakage*, *GUI phishing*, and *repackaging*. In addition, we position *attack-based* solutions based on the deployment classification, specifically, into *workstation*, *mobile*, and *hybrid* architecture based solutions.

2.6.1 Workstation-Based Solutions

The following solutions constitute *workstation-based* solutions, in which the authors target large-scale deployment.

ICCDetector [83] is a machine learning system to detect ICC-based (Android Inter-Component Communication) malicious apps. The authors adopt the following approach for the proposed system: (1) They extract ICC features for the Android app using a preexisting tool called EPICC. (2) They apply feature selections using Correlation-based Feature Selection (CFS). (3) Finally, the selected ICC features using EPICC are normalized in feature vectors to be input to a binary classifier (SVM, Decision Tree, Random Forest). The feature vector stores the occurrence number of a given feature. In the detection phase, the ICCDetector system uses the trained model to detect ICC-based malware. Using ICC for malware detection, ICCDetector aims to fill the detection of ICC-based malware gap that relies on ICC malicious payload.

AnDarwin [73, 84] detects similar apps that are written by the same developer as well as different developers. AnDarwin consists of four stages: (1) AnDarwin extracts similar vectors by computing an undirected PDG (Program Dependence Graph) of each method in the app using only data dependencies for the edges. (2) AnDarwin finds similar code segments by clustering all the vectors of all apps. It identifies code clones by finding near-neighbors of vectors using Locality Sensitive Hashing (LSH). (3) AnDarwin eliminates library code based on the frequency of the clusters. (4) AnDarwin detects similar apps (full and partial detection) by computing the pairwise similarity between all the sets using LSH techniques, specifically MinHash. Because AnDarwin is based on PDG for extracting semantics vectors, it is less scalable with respect to analysis time. Using 75 threads, AnDarwin takes 4.47 days to extract semantic vectors (first stage) from 265,359 apps.

FlowDroid [85] performs a context, flow, object, and field-sensitive static taint analysis on Android apps. It models Android app's lifecycle states and handles taint propagation due to callbacks and User Interface (UI) objects. It also utilizes SuSi [86], a machine-learning-based technique, to automatically identify sources and sinks of sensitive information in an Android API. FlowDroid achieves 86% precision and 93% recall, which represent better results than two commercial tools: AppScan and Fortify SCA. However, the system uses high-complexity tools to process Apk files, such as Soot [87] and Dexpler [88]. Also, it does not track data flows across different app components that communicate using Android ICC.

MassVet [67] is designed for vetting apps at a massive scale, without knowing what malware looks like and how it behaves. It runs a DiffCom analysis of the submitted app against the whole market app. It compares a submitted app with all apps already on the market by focusing on the difference between those sharing a similar UI structure, which is known to be primarily preserved during repackaging. It also performs an intersection analysis to compare the new apps against existing apps with different view structures and signed by various certificates. The aim is to inspect their common parts to identify suspicious code segments (at the method level).

2.6.2 Mobile-Based Solutions

In the following, we present *mobile-based* solutions that are meant to be deployed on smart mobile devices.

Aurasium [89] automatically repackages an application to attach a user-level policy enforcement code. The role of this code is to monitor any security violations, such as sending SMS to premium charging numbers. If such a case occurs, Aurasium displays the destination number and the SMS content, so the user can confirm or deny the operation.

AppGuard [90] is a policy enforcement system, which provides the user with the ability to revoke permissions after app-installation time. It takes an untrusted app and user-defined security policies as input and inserts a security monitor API into the untrusted app by repackaging the Apk. Security policies may specify restrictions on method invocations as well as secrecy requirements. This requires the identification of relevant methods at the API level for which such checks are required.

Patronus [91] is a device-based Intrusion Prevention System (IPS). The latter is a rule (policy) based system, in which there is a policy database that defines what malicious behaviors are. Patronus comprises two main parts: the client-side app and the server-side service. Both parts are on Android devices. The client is a simple Android app; however, the server-side only updates the policy files in the Android OS. Therefore, there is no need to change the actual Android OS; it only requires to inject Patronus into the system to build a hook and capture app's behaviors. Patronus needs a special privilege to insert such files, which prevents a large-scale deployment. Also, injecting files into Android OS triggers some security concerns on these files. Finally, the policy-based detection system is limited to the expressiveness of the rules in the database.

2.6.3 Hybrid Solutions

This section depicts solutions that leverage *hybrid* architecture.

XDroid [92] is proposed as a risk assessment tool and a user alert generator. XDroid has two components: (1) XDroid client, which monitors app's behaviors, such as Android API calls, and timestamps these events. These events are sent continuously to an XDroid Server, where risk assessment, user profiling, and alert customization services leverage time-series of events to assess the suspicious behavior of a given app. To do so, the authors propose the use of Hidden Markov Model (HMM), which is first trained on malicious and benign behavior services. Afterward, the trained HMM Model is deployed to server. However, XDroid needs a special privilege in Android devices to be able to log app's behaviors; this could prevent this technique from having large-scale deployments since devices need to be rooted to track app behaviors properly.

DroidEagle [68] uses the layout resources within an app to detect visually similar apps, a common characteristic in repackaged apps and phishing malware. To discover visually similar apps, DroidEagle consists of two sub-systems: RepoEagle and HostEagle. RepoEagle performs large-scale detection on apps repositories (e.g., apps markets), and HostEagle is a lightweight mobile app that can help users to detect visually similar Android apps quickly upon download. The reported performance results show that, within 3 h, RepoEagle can detect 1298 visually similar apps from 99,626 apps in a repository.

2.6.4 Discussions

Table 2.5 shows the comparison between *attack-based* solutions using *feature selection* and *detection strategy* criteria. As shown in Table 2.5, most *Workstation-based* solutions are based on static analysis. In addition to code static analysis, some solutions statically analyzed the resources of the Apk file, for instance, MassVet [67] leverages only the GUI XML resources files in the Apk and ignore the rest of the content. Furthermore, *Workstation-based* solutions tend to be resilient against known obfuscation techniques since such technique mainly target bytecode, while these solutions do not rely only on bytecode to detect Android malicious apps. On the other hand, *mobile-based* solutions [89–91] rely only on behavioral features of app runtime on the end-users mobile devices. These features are highly resilient to obfuscation and adapt to OS changes since they are collected from executions

Table 2.5 Qualitative comparison of specific malware attack solutions

Solution	Feature selection			Detection strategy			
	Type	Resiliency	Adapt	Approach	Scope	Mode	Response
ICCDetector [83]	Static	High	Low	Learning	Known	Supervised	–
Andarwin [84]	Static	High	High	Signature	Known	–	–
Flowdroid [85]	Static	High	High	Prior-knowledge	Known, zero-day	–	–
MassVet [67]	Static	High	High	Signature	Known, zero-day	–	–
Aurasium [89]	Behavioral	Highest	Highest	Prior-knowledge	Known, zero-day	–	Passive
AppGuard [90]	Behavioral	Highest	Highest	Prior-knowledge	Known, zero-day	–	Passive
Patronus [91]	Behavioral	Highest	Highest	Prior-knowledge	Known, zero-day	–	Active
XDroid [92]	Behavioral	Highest	Highest	Learning	Known	Unsupervised	Active
Droideagle [68]	Static	High	High	Signature	Known, zero-day	–	Passive

Table 2.6 Quantitative comparison of specific malware attack solutions

Solution	APK prepro-cessing complex-ity	Feature extraction complex-ity	Model genera-tion computa-tion	Detection rate	Detection time	Dataset
ICCDetector [83]	High	Low	Low	Acc = 97.4%	About 40 s	Ben = 12k, Mal = 5.3k
Andarwin [84]	High	High	Lowest	Found = 36k Rebranded App	10 h	Apps = 266k
Flowdroid [85]	Medium	High	Lowest	F1 = 90%	16 s	Ben = 1k, Mal = 0.5k
MassVet [67]	Medium	Medium	Lowest	Acc = 72%	10 s	Apps = 1.2 Million
Aurasium [89]	High	Lowest	Lowest	Success = 99.6%	–	Ben = 3.5k, Mal = 1.3k
AppGuard [90]	High	Lowest	Lowest	–	About 20 s	Apps = 250k
Patronus [91]	Low	Lowest	Lowest	F1 = 69–92%	–	Ben = 0.5k, Mal = 0.3k
XDroid [92]	Low	Low	Low	F1 = 83%	–	Benn = 0.7k, Mal=5.6k
Droideagle [68]	Low	Low	Lowest	Similar = 1.3K apps	Average = 62 s	Apps = 100k

of the malicious apps. Both Aurasium [89] and AppGuard [90] provide a passive notification to end-users; Patronus [91] goes a step further by blocking the detected malware. Also, *mobile-based* solutions rely on policy rules to prevent malicious behaviors. *Hybrid* solutions employ behavioral analysis as in XDroid [92] and static analysis as proposed in Droideagle [68]. The latter is resilient to code obfuscation because it relies on GUI similarity and does not consider the bytecode.

Table 2.6 shows a quantitative comparison between *attack-based* solutions. *Apk preprocessing complexity* varies among the solutions; it is medium in some solutions [67, 85] and high in other solutions [83, 84] that use static analysis because the preprocessing of Apks. In case of Aurasium [89] and AppGuard [90] the *Apk preprocessing complexity* is high because of the Apk's repackaging complexity to inject the monitoring Apk hooks. It is low for XDroid [92] and Droideagle [68] because the first [92] relies mainly on runtime traces and the second [68] uses Apk GUI resources for its graphical signatures. Finally, *attack-based* solutions have very low complexity for model generation because most of these models are based on policy rules databases.

Similarly to *generic* class, the *attack-based* solutions target Android malware detection. However, these solutions target specific malicious attacks, in contrast to *general solutions*, which target malware in general. For instance, ICCDetector [83] aims to detect malware that employ Android Inter-Components Communication

Table 2.7 Classifications of specific malware attack solutions

Solution	Layer	Architecture	Physical components	Target application
ICCDetector [83]	App layer	Workstation	PC	ICC abuse
Andarwin [84]	App layer	Workstation	Server	Find repackaging
Flowdroid [85]	App layer	Workstation	PC	Sensitive data leakage
MassVet [67]	App layer	Workstation	PC	Find GUI repackaging
Aurasium [89]	App, framework layers	Mobile	Mobile/IoT devices	Policy enforcement
AppGuard [90]	Linux kernel	Mobile	Mobile/IoT devices	Policy enforcement
Patronus [91]	App, framework layers	Mobile	Mobile devices	Mobile IPS
XDroid [92]	App, framework layers	Hybrid	Mobile/IoT devices, server	Risk assessment
Droideagle [68]	App, framework layers	Hybrid	Mobile devices, server	GUI phishing and repackaging

(ICC) to their attack. Another example, Flowdroid [85] helps detecting sensitive data leakages, which is a known pattern of malicious apps. Table 2.7 shows that *workstation-based* solutions [67, 83–85] rely on app layers of Android stack; the other solutions (*mobile-based* and *hybrid*) are implemented across app and framework layers. The exception here is AppGuard [90], which is implemented on Linux kernel layer.

2.7 Android Malware Detection Helpers

In this section, we present systems that help enhance Android malware detection systems. These solutions do not provide a malware detection system, but they are used as tools to enhance the malware detection in term of accuracy and runtime performance. The following solutions are lab tools that could be leveraged for malware detection.

The authors in [93] analyze the Android framework statically. The authors propose a top-down approach in their analysis by taking the source code of the Android framework as input. For this purpose, the authors face different challenges: (1) The Android framework layer is different from the application layer, so the existing analysis tools and techniques cannot be used to analyze the framework layer. (2) The framework services may be queried simultaneously from multiple apps. Thus, the framework layer uses various multi-threading mechanisms. (3) It is unclear what resources are protected by Android permissions. The authors use their analysis insights to develop the Axplorer tool [93] for the analysis of the Android framework layer. To demonstrate the effectiveness of their security analysis, the authors conduct a permission API mapping (map a given permission to the related Android Framework API). In comparison to previous research, they achieve a more precise analysis. Finally, they propose a permission locality security concept to measure permission coverage overlap of Android APIs.

DroidChameleon [94] is a tool to evaluate the robustness and resiliency of anti-malware solutions against the state-of-the-art obfuscation techniques. Droid-Chameleon provides a set of obfuscation techniques to be tested on targeted malware detection systems. DroidChameleon supports three levels of obfuscation techniques: (1) Trivial obfuscations, that do not change the bytecode but only repackages API file or disassembling and Reassembling the Dex file. (2) Attacks detectable by static analysis obfuscations that make a change to the bytecode in one or multiple ways such as identifier renaming, code reordering, junk code insertion, which could be detected using static analysis. (3) Attacks that could not be identified using static analysis such as reflection and bytecode encryption. The authors conduct a systematic evaluation of existing anti-malware products against various obfuscation techniques; this is a necessary evaluation to measure the resiliency of existing malware detection systems.

2.7.1   Discussions

Demystifying [93] proposes an in-depth analysis of the Android framework. As an application for this analysis, they reevaluate Android permission mapping to the actual Android assets with high precision. Having such precise mapping helps in malware detection. Given an Android app the detection system could map permissions to a more granular view by using Android assets. On the other hand, the authors of DroidChameleon [94] propose a tool that provides a set of obfuscation techniques to be applied to Android apps. DroidChameleon is a valuable tool that helps enhancing malware detection system by evaluating these systems on obfuscated malware and benign sample using DroidChameleon. In a security lab environment, one could ensure that the detection system is resilient to Droid-Chameleon obfuscation techniques. A simple classification of Android malware detection helpers is depicted in Table 2.8. In this book, we use DroidChameleon

Table 2.8 Classification of android malware detection helpers

Solution	Layer	Architecture	Physical components	Target application
Demystifying [93]	Framework layer	Workstation	Server	Android framework analytics tool
DroidChameleon [94]	App layer	Workstation	PC	Obfuscation tool

obfuscation tool to build an Android obfuscation dataset that will be used for the evaluation of elaborated systems in the next chapters.

2.8 Summary

In this chapter, we reviewed selected prominent Android malware detection solutions and proposed a taxonomy to classify them. This taxonomy allows to carry out an in-depth comparative study between the proposals of the same category. By proposing a generic functional framework for Android malware detection, we defined the performance criteria used for our comparative study. Through the study of different Android malware detection systems, we define the advantages and disadvantages of each system approach. Moreover, we illustrated the situations where each approach exhibits good performance.

Android detection systems with high detection accuracy incur a long processing time for complex feature extraction, which implies a high detection latency and the use of more computation resources. On the other hand, detection systems with less sophisticated feature extraction achieve a moderate detection accuracy with the advantage of low detection latency and small resources needed, which allows such systems to function properly on mobile and IoT devices. To this end, an appealing future work lies in achieving simultaneously high detection results and a low detection latency with the minimum computation resources to fit all scales of devices.

In the next chapter, we propose a solution for Android malware clustering using static analysis features and graph partitioning algorithms. According to the proposed taxonomy discussed in this chapter, we position this solution as (1) general-attack oriented because it targets the detection of all types of Android malware. Also, we consider it as (2) workstation oriented due to the level of resources it requires in the deployment. Finally, it is an application layer solution because it considers only Android apps.

References

1. Open Handset Alliance, https://www.openhandsetalliance.com/android_overview.html. Accessed Dec 2016
2. The Android Native Development Kit (NDK), https://developer.android.com/ndk/index.html. Accessed Jan 2016
3. Android Platform Architecture, https://developer.android.com/guide/platform/index.html. Accessed March 2017
4. J. Oberheide, C. Miller, Dissecting the android bouncer, in *SummerCon2012, New York* (2012)
5. Appchine Market, http://www.appchina.com/. Accessed March 2017
6. Mumayi Market, http://www.mumayi.com/. Accessed Jan 2017

7. Beware! New Android Malware Infected 2 Million Google Play Store Users, http://thehackernews.com/2017/04/android-malware-playstore.html. Accessed April 2017
8. HummingBad Android Malware Found in 20 Google Play Store Apps, https://www.bleepingcomputer.com/news/security/hummingbad-android-malware-found-in-20-google-play-store-apps/. Accessed Jan 2017
9. List: 44 Android apps infected with malware made their way to the Google Play store, http://clark.com/technology/google-play-malware-app-hummingbad. Accessed Dec 2017
10. A.P. Felt, E. Ha, S. Egelman, A. Haney, E. Chin, D.A. Wagner, Android permissions: user attention, comprehension, and behavior, in *Symposium On Usable Privacy and Security, SOUPS '12, Washington, DC, USA*, 11–13 July 2012, p. 3
11. A.P. Felt, E. Chin, S. Hanna, D. Song, D.A. Wagner, Android permissions demystified, in *Proceedings of the 18th ACM Conference on Computer and Communications Security, CCS 2011, Chicago, Illinois, USA*, 17–21 Oct 2011, pp. 627–638
12. E.B. Karbab, M. Debbabi, S. Alrabaee, D. Mouheb, Dysign: dynamic fingerprinting for the automatic detection of android malware. CoRR, abs/1702.05699 (2017)
13. E.B. Karbab, M. Debbabi, S. Alrabaee, D. Mouheb, Dysign: dynamic fingerprinting for the automatic detection of android malware, in *11th International Conference on Malicious and Unwanted Software, MALWARE 2016, Fajardo, PR, USA*, 18–21 Oct 2016, pp. 139–146
14. D. Arp, M. Spreitzenbarth, M. Hubner, H. Gascon, K. Rieck, DREBIN: effective and explainable detection of android malware in your pocket, in *21st Annual Network and Distributed System Security Symposium, NDSS 2014, San Diego, California, USA*, 23–26 Feb 2014
15. Y. Feng, S. Anand, I. Dillig, A. Aiken, Apposcopy: semantics-based detection of android malware through static analysis, in *Proceedings of the 22nd ACM SIGSOFT International Symposium on Foundations of Software Engineering, (FSE-22), Hong Kong, China*, 16–22 Nov 2014, pp. 576–587
16. A.I. Ali-Gombe, I. Ahmed, G.G. Richard III, V. Roussev, Aspectdroid: android app analysis system, in *Proceedings of the Sixth ACM on Conference on Data and Application Security and Privacy, CODASPY 2016, New Orleans, LA, USA*, 9–11 March 2016, pp. 145–147
17. G. Canfora, E. Medvet, F. Mercaldo, C.A. Visaggio, Acquiring and analyzing app metrics for effective mobile malware detection, in *Proceedings of the 2016 ACM on International Workshop on Security and Privacy Analytics, IWSPA@CODASPY 2016, New Orleans, LA, USA*, 11 March 2016, pp. 50–57
18. S. Bhandari, R. Gupta, V. Laxmi, M.S. Gaur, A. Zemmari, M. Anikeev, DRACO: droid analyst combo an android malware analysis framework, in *Proceedings of the 8th International Conference on Security of Information and Networks, SIN 2015, Sochi, Russian Federation*, 8–10 Sept 2015, pp. 283–289
19. M. Zhang, Y. Duan, H. Yin, Z. Zhao, Semantics-aware android malware classification using weighted contextual API dependency graphs, in *Proceedings of the 2014 ACM SIGSAC Conference on Computer and Communications Security, Scottsdale, AZ, USA* 3–7 Nov 2014, pp. 1105–1116
20. E.B. Karbab, M. Debbabi, D. Mouheb, Fingerprinting Android packaging: generating DNAs for malware detection. Digit. Investig. **18**, S33–S45 (2016)
21. W. Yang, J. Li, Y. Zhang, Y. Li, J. Shu, D. Gu, Apklancet: tumor payload diagnosis and purification for android applications, in *9th ACM Symposium on Information, Computer and Communications Security, ASIA CCS '14, Kyoto, Japan*, 03–06 June 2014, pp. 483–494
22. Y. Zhongyang, Z. Xin, B. Mao, L. Xie, Droidalarm: an all-sided static analysis tool for android privilege-escalation malware, in *8th ACM Symposium on Information, Computer and Communications Security, ASIA CCS '13, Hangzhou, China*, 08–10 May 2013, pp. 353–358
23. B. Sanz, I. Santos, X. Ugarte-Pedrero, C. Laorden, J. Nieves, P.G. Bringas, Anomaly detection using string analysis for android malware detection, in *International Joint Conference SOCO'13-CISIS'13-ICEUTE'13 -Proceedings, Salamanca, Spain*, 11–13 Sept 2013, pp. 469–478

24. T. Kim, B. Kang, M. Rho, S. Sezer, E.G. Im, A multimodal deep learning method for android malware detection using various features. IEEE Trans. Inf. Forensics Secur. **14**(3), 773–788 (2019)
25. L. Onwuzurike, E. Mariconti, P. Andriotis, E.D. Cristofaro, G.J. Ross, G. Stringhini, Mamadroid: detecting android malware by building Markov chains of behavioral models (extended version). ACM Trans. Priv. Secur. **22**(2), 14:1–14:34 (2019)
26. K. Xu, Y. Li, R.H. Deng, K. Chen, Deeprefiner: multi-layer android malware detection system applying deep neural networks, in *2018 IEEE European Symposium on Security and Privacy, EuroS&P 2018, London, United Kingdom*, 24–26 April 2018, pp. 473–487
27. K. Xu, Y. Li, R.H. Deng, K. Chen, J. Xu, Droidevolver: self-evolving android malware detection system, in *IEEE European Symposium on Security and Privacy, EuroS&P 2019, Stockholm, Sweden*, 17–19 June 2019, pp. 47–62
28. J. Allen, M. Landen, S. Chaba, Y. Ji, S.P.H. Chung, W. Lee, Improving accuracy of android malware detection with lightweight contextual awareness, in *Proceedings of the 34th Annual Computer Security Applications Conference, ACSAC 2018, San Juan, PR, USA*, 03–07 Dec 2018, pp. 210–221
29. K.O. Elish, X. Shu, D.D. Yao, B.G. Ryder, X. Jiang, Profiling user-trigger dependence for android malware detection. Comput. Secur. **49**, 255–273 (2015)
30. F. Idrees, M. Rajarajan, M. Conti, T.M. Chen, Y. Rahulamathavan, Pindroid: a novel android malware detection system using ensemble learning methods. Comput. Secur. **68**, 36–46 (2017)
31. P. Burnap, R. French, F. Turner, K. Jones, Malware classification using self organising feature maps and machine activity data. Comput. Secur. **73**, 399–410 (2018)
32. S. Badhani, S.K. Muttoo, Cendroid - a cluster-ensemble classifier for detecting malicious android applications. Comput. Secur. **85**, 25–40 (2019)
33. P. Faruki, A. Bharmal, V. Laxmi, V. Ganmoor, M.S. Gaur, M. Conti, M. Rajarajan, Android security: a survey of issues, malware penetration, and defenses. IEEE Commun. Surv. Tutor. **17**(2), 998–1022 (2015)
34. P. Faruki, V. Ganmoor, V. Laxmi, M.S. Gaur, A. Bharmal, Androsimilar: robust statistical feature signature for android malware detection, in *The 6th International Conference on Security of Information and Networks, SIN '13, Aksaray, Turkey*, 26–28 Nov 2013, pp. 152–159
35. E. Chin, A.P. Felt, K. Greenwood, D.A. Wagner, Analyzing inter-application communication in android, in *Proceedings of the 9th International Conference on Mobile Systems, Applications, and Services (MobiSys 2011), Bethesda, MD, USA*, 28 June–01 July 2011, pp. 239–252
36. A.P. Fuchs, A. Chaudhuri, J.S. Foster, Scandroid: automated security certification of android (2009). https://www.cs.umd.edu/~avik/papers/scandroidascaa.pdf
37. B.P. Sarma, N. Li, C.S. Gates, R. Potharaju, C. Nita-Rotaru, I. Molloy, Android permissions: a perspective combining risks and benefits, in *17th ACM Symposium on Access Control Models and Technologies, SACMAT '12, Newark, NJ, USA*, 20–22 June 2012, pp. 13–22
38. D. Barrera, H.G. Kayacik, P.C. van Oorschot, A. Somayaji, A methodology for empirical analysis of permission-based security models and its application to android, in *Proceedings of the 17th ACM Conference on Computer and Communications Security, CCS 2010, Chicago, Illinois, USA*, 4–8 Oct 2010, pp. 73–84
39. W. Enck, M. Ongtang, P.D. McDaniel, On lightweight mobile phone application certification, in *Proceedings of the 2009 ACM Conference on Computer and Communications Security, CCS 2009, Chicago, Illinois, USA*, 9–13 Nov 2009, pp. 235–245
40. M.C. Grace, Y. Zhou, Q. Zhang, S. Zou, X. Jiang, Riskranker: scalable and accurate zero-day android malware detection, in *The 10th International Conference on Mobile Systems, Applications, and Services, MobiSys'12, Ambleside, United Kingdom*, 25–29 June 2012, pp. 281–294
41. J. Kim, Y. Yoon, K. Yi, J. Shin, S. Center, (POSTER) ScanDal: static analyzer for detecting privacy leaks in android applications. IEEE Secur. Priv. **12**(1), 1–10 (2012)

42. Y. Zhang, M. Yang, B. Xu, Z. Yang, G. Gu, P. Ning, X.S. Wang, B. Zang, Vetting undesirable behaviors in android apps with permission use analysis, in *2013 ACM SIGSAC Conference on Computer and Communications Security, CCS'13, Berlin, Germany*, 4–8 Nov 2013, pp. 611–622

43. B. Amos, H.A. Turner, J. White, Applying machine learning classifiers to dynamic android malware detection at scale, in *2013 9th International Wireless Communications and Mobile Computing Conference, IWCMC 2013, Sardinia, Italy*, 1–5 July 2013, pp. 1666–1671

44. T. Wei, C. Mao, A.B. Jeng, H. Lee, H. Wang, D. Wu, Android malware detection via a latent network behavior analysis, in *11th IEEE International Conference on Trust, Security and Privacy in Computing and Communications, TrustCom 2012, Liverpool, United Kingdom*, 25–27 June 2012, pp. 1251–1258

45. J. Huang, X. Zhang, L. Tan, P. Wang, B. Liang, Asdroid: detecting stealthy behaviors in android applications by user interface and program behavior contradiction, in *36th International Conference on Software Engineering, ICSE '14, Hyderabad, India*, 31 May–07 June 2014, pp. 1036–1046

46. A. Saracino, D. Sgandurra, G. Dini, F. Martinelli, MADAM: effective and efficient behavior-based android malware detection and prevention. IEEE Trans. Dependable Sec. Comput. **15**(1), 83–97 (2018)

47. W. Enck, P. Gilbert, B. Chun, L.P. Cox, J. Jung, P.D. McDaniel, A. Sheth, Taintdroid: an information flow tracking system for real-time privacy monitoring on smartphones. Commun. ACM **57**(3), 99–106 (2014)

48. I. Burguera, U. Zurutuza, S. Nadjm-Tehrani, Crowdroid: behavior-based malware detection system for android, in *SPSM'11, Proceedings of the 1st ACM Workshop Security and Privacy in Smartphones and Mobile Devices, Co-located with CCS 2011, Chicago, IL, USA*, 17 Oct 2011, pp. 15–26

49. K.O. Elish, D. Yao, B.G. Ryder, User-centric dependence analysis for identifying malicious mobile apps, in *2012 IEEE Security and Privacy Workshops, SP Workshops 2012* (2016)

50. A. Shabtai, U. Kanonov, Y. Elovici, C. Glezer, Y. Weiss, "andromaly": a behavioral malware detection framework for android devices. J. Intell. Inf. Syst. **38**(1), 161–190 (2012)

51. A. Reina, A. Fattori, L. Cavallaro, A system call-centric analysis and stimulation technique to automatically reconstruct android malware behaviors, in *EuroSec*, April 2013

52. D. Damopoulos, G. Kambourakis, G. Portokalidis, The best of both worlds: a framework for the synergistic operation of host and cloud anomaly-based IDS for smartphones, in *Proceedings of the Seventh European Workshop on System Security, EuroSec 2014, Amsterdam, The Netherlands*, 13 April 2014, pp. 6:1–6:6

53. M. Spreitzenbarth, F.C. Freiling, F. Echtler, T. Schreck, J. Hoffmann, Mobile-sandbox: having a deeper look into android applications, in *Proceedings of the 28th Annual ACM Symposium on Applied Computing, SAC '13, Coimbra, Portugal*, 18–22 March 2013, pp. 1808–1815

54. M. Lindorfer, M. Neugschwandtner, L. Weichselbaum, Y. Fratantonio, V. van der Veen, C. Platzer, ANDRUBIS - 1, 000, 000 apps later: a view on current android malware behaviors, in *Third International Workshop on Building Analysis Datasets and Gathering Experience Returns for Security, BADGERS@ESORICS 2014, Wroclaw, Poland*, 11 Sept 2014, pp. 3–17

55. T. Vidas, J. Tan, J. Nahata, C.L. Tan, N. Christin, P. Tague, A5: automated analysis of adversarial android applications, in *Proceedings of the 4th ACM Workshop on Security and Privacy in Smartphones & Mobile Devices, SPSM@CCS 2014, Scottsdale, AZ, USA*, 03–07 Nov 2014, pp. 39–50

56. Y. Zhou, Z. Wang, W. Zhou, X. Jiang, Hey, you, get off of my market: Detecting malicious apps in official and alternative android markets, in *19th Annual Network and Distributed System Security Symposium, NDSS 2012, San Diego, California, USA*, 5–8 Feb 2012

57. F. Martinelli, F. Mercaldo, A. Saracino, BRIDEMAID: an hybrid tool for accurate detection of android malware, in *Proceedings of the 2017 ACM on Asia Conference on Computer and Communications Security, AsiaCCS 2017, Abu Dhabi, United Arab Emirates*, 2–6 April 2017, pp. 899–901

58. J. Jang, H. Kang, J. Woo, A. Mohaisen, H.K. Kim, Andro-dumpsys: anti-malware system based on the similarity of malware creator and malware centric information. Comput. Secur. **58**, 125–138 (2016)

59. A.I. Ali-Gombe, B. Saltaformaggio, J. Ramanujam, D. Xu, G.G. Richard III, Toward a more dependable hybrid analysis of android malware using aspect-oriented programming. Comput. Secur. **73**, 235–248 (2018)

60. A.I. Ali-Gombe, I. Ahmed, G.G. Richard III, V. Roussev, Opseq: android malware fingerprinting, in *Proceedings of the 5th Program Protection and Reverse Engineering Workshop, PPREW@ACSAC, Los Angeles, CA, USA*, 8 Dec 2015, pp. 7:1–7:12

61. L. Deshotels, V. Notani, A. Lakhotia, Droidlegacy: automated familial classification of android malware, in *Proceedings of the 3rd ACM SIGPLAN Program Protection and Reverse Engineering Workshop 2014, PPREW 2014, San Diego, CA, USA*, 25 Jan 2014, pp. 3:1–3:12 (2014)

62. J. Lee, S. Lee, H. Lee, Screening smartphone applications using malware family signatures. Comput. Secur. **52**, 234–249 (2015)

63. J. Kim, T. Kim, E.G. Im, Structural information based malicious app similarity calculation and clustering, in *Proceedings of the 2015 Conference on Research in Adaptive and Convergent Systems, RACS 2015, Prague, Czech Republic*, 9–12 Oct 2015, pp. 314–318

64. G. Suarez-Tangil, J.E. Tapiador, P. Peris-Lopez, J.B. Alís, Dendroid: a text mining approach to analyzing and classifying code structures in android malware families. Expert Syst. Appl. **41**(4), 1104–1117 (2014)

65. Y. Lin, Y. Lai, C. Chen, H. Tsai, Identifying android malicious repackaged applications by thread-grained system call sequences. Comput. Secur. **39**, 340–350 (2013)

66. P. Faruki, V. Laxmi, A. Bharmal, M.S. Gaur, V. Ganmoor, Androsimilar: robust signature for detecting variants of android malware. J. Inf. Sec. Appl. **22**, 66–80 (2015)

67. K. Chen, P. Wang, Y. Lee, X. Wang, N. Zhang, H. Huang, W. Zou, P. Liu, Finding unknown malice in 10 seconds: mass vetting for new threats at the google-play scale, in *24th USENIX Security Symposium, USENIX Security 15, Washington, D.C., USA*, 12–14 Aug 2015, pp. 659–674

68. M. Sun, M. Li, J.C.S. Lui, Droideagle: seamless detection of visually similar android apps, in *Proceedings of the 8th ACM Conference on Security & Privacy in Wireless and Mobile Networks, New York, NY, USA*, 22–26 June 2015, pp. 9:1–9:12

69. W. Zhou, Y. Zhou, X. Jiang, P. Ning, Detecting repackaged smartphone applications in third-party android marketplaces, in *Second ACM Conference on Data and Application Security and Privacy, CODASPY 2012, San Antonio, TX, USA*, 7–9 Feb 2012, pp. 317–326

70. S. Hanna, L. Huang, E.X. Wu, S. Li, C. Chen, D. Song, Juxtapp: a scalable system for detecting code reuse among android applications, in *Detection of Intrusions and Malware, and Vulnerability Assessment - 9th International Conference, DIMVA 2012, Heraklion, Crete, Greece, Revised Selected Papers*, 26–27 July 2012, pp. 62–81

71. J. Crussell, C. Gibler, H. Chen, Attack of the clones: detecting cloned applications on android markets, in *Computer Security - ESORICS 2012 Proceedings - 17th European Symposium on Research in Computer Security, Pisa, Italy*, 10–12 Sept 2012, pp. 37–54

72. W. Zhou, Y. Zhou, M.C. Grace, X. Jiang, S. Zou, Fast, scalable detection of "piggybacked" mobile applications, in *Third ACM Conference on Data and Application Security and Privacy, CODASPY'13, San Antonio, TX, USA*, 18–20 Feb 2013, pp. 185–196

73. J. Crussell, C. Gibler, H. Chen, Andarwin: scalable detection of semantically similar android applications, in *Computer Security - ESORICS 2013 Proceedings - 18th European Symposium on Research in Computer Security, Egham, UK*, 9–13 Sept 2013, pp. 182–199

74. E.B. Karbab, M. Debbabi, A. Derhab, D. Mouheb, Cypider: building community-based cyber-defense infrastructure for android malware detection, in *Proceedings of the 32nd Annual Conference on Computer Security Applications, ACSAC 2016, Los Angeles, CA, USA*, 5–9 Dec 2016, pp. 348–362

75. K. Tian, D.D. Yao, B.G. Ryder, G. Tan, G. Peng, Detection of repackaged android malware with code-heterogeneity features. IEEE Trans. Dependable Secure Comput. (2017). https://doi.org/10.1109/TDSC.2017.2745575

76. M. Fan, J. Liu, X. Luo, K. Chen, Z. Tian, Q. Zheng, T. Liu, Android malware familial classification and representative sample selection via frequent subgraph analysis. IEEE Trans. Inf. Forensics Secur. **13**(8), 1890–1905 (2018)

77. E. Mariconti, L. Onwuzurike, P. Andriotis, E.D. Cristofaro, G.J. Ross, G. Stringhini, Mamadroid: detecting android malware by building Markov chains of behavioral models, in *24th Annual Network and Distributed System Security Symposium, NDSS 2017, San Diego, California, USA*, 26 Feb–1 March 2017

78. L. Onwuzurike, E. Mariconti, P. Andriotis, E.D. Cristofaro, G.J. Ross, G. Stringhini, Mamadroid: detecting android malware by building Markov chains of behavioral models (extended version). ACM Trans. Priv. Secur. **22**(2), 14:1–14:34 (2019)

79. S. Chen, M. Xue, Z. Tang, L. Xu, H. Zhu, Stormdroid: a streaminglized machine learning-based system for detecting android malware, in *Proceedings of the 11th ACM on Asia Conference on Computer and Communications Security, AsiaCCS 2016, Xi'an, China*, 30 May–3 June 2016, pp. 377–388

80. H.M.J. Almohri, D.D. Yao, D.G. Kafura, Droidbarrier: know what is executing on your android, in *Fourth ACM Conference on Data and Application Security and Privacy, CODASPY'14, San Antonio, TX, USA*, 03–05 March 2014, pp. 257–264

81. J. Sahs, L. Khan, A machine learning approach to android malware detection, in *2012 European Intelligence and Security Informatics Conference, EISIC 2012, Odense, Denmark*, 22–24 Aug 2012, pp. 141–147

82. M. Sun, X. Li, J.C.S. Lui, R.T.B. Ma, Z. Liang, Monet: a user-oriented behavior-based malware variants detection system for android. IEEE Trans. Inf. Forensics Secur. **12**(5), 1103–1112 (2017)

83. K. Xu, Y. Li, R.H. Deng, Iccdetector: Icc-based malware detection on android. IEEE Trans. Inf. Forensics Secur. **11**(6), 1252–1264 (2016)

84. J. Crussell, C. Gibler, H. Chen, Andarwin: scalable detection of android application clones based on semantics. IEEE Trans. Mob. Comput. **14**(10), 2007–2019 (2015)

85. S. Arzt, S. Rasthofer, C. Fritz, E. Bodden, A. Bartel, J. Klein, Y.L. Traon, D. Octeau, P.D. McDaniel, Flowdroid: precise context, flow, field, object-sensitive and lifecycle-aware taint analysis for android apps, in *ACM SIGPLAN Conference on Programming Language Design and Implementation, PLDI '14, Edinburgh, United Kingdom*, 09–11 June 2014, pp. 259–269

86. S. Rasthofer, S. Arzt, E. Bodden, A machine-learning approach for classifying and categorizing android sources and sinks, in *21st Annual Network and Distributed System Security Symposium, NDSS 2014, San Diego, California, USA*, 23–26 Feb 2014

87. P. Lam, E. Bodden, O. Lhoták, L. Hendren, The Soot framework for Java program analysis: a retrospective, in *Cetus Users and Compiler Infrastructure Workshop (CETUS 2011)*, vol. 15 (2011), p. 35

88. A. Bartel, J. Klein, Y.L. Traon, M. Monperrus, Dexpler: converting android Dalvik bytecode to Jimple for static analysis with soot, in *Proceedings of the ACM SIGPLAN International Workshop on State of the Art in Java Program analysis, SOAP 2012, Beijing, China*, 14 June 2012, pp. 27–38

89. R. Xu, H. Saïdi, R.J. Anderson, Aurasium: practical policy enforcement for android applications, in *Proceedings of the 21th USENIX Security Symposium, Bellevue, WA, USA*, 8–10 Aug 2012, pp. 539–552

90. M. Backes, S. Gerling, C. Hammer, M. Maffei, P. von Styp-Rekowsky, Appguard - fine-grained policy enforcement for untrusted android applications, in *Data Privacy Management and Autonomous Spontaneous Security - 8th International Workshop, DPM 2013, and 6th International Workshop, SETOP 2013, Egham, UK, Revised Selected Papers*, 12–13 Sept 2013, pp. 213–231

91. M. Sun, M. Zheng, J.C.S. Lui, X. Jiang, Design and implementation of an android host-based intrusion prevention system, in *Proceedings of the 30th Annual Computer Security Applications Conference, ACSAC 2014, New Orleans, LA, USA*, 8–12 Dec 2014, pp. 226–235
92. B. Rashidi, C.J. Fung, E. Bertino, Android resource usage risk assessment using hidden Markov model and online learning. Comput. Secur. **65**, 90–107 (2017)
93. M. Backes, S. Bugiel, E. Derr, P.D. McDaniel, D. Octeau, S. Weisgerber, On demystifying the android application framework: re-visiting android permission specification analysis, in *25th USENIX Security Symposium, USENIX Security 16, Austin, TX, USA*, 10–12 Aug 2016, pp. 1101–1118
94. V. Rastogi, Y. Chen, X. Jiang, Droidchameleon: evaluating android anti-malware against transformation attacks, in *8th ACM Symposium on Information, Computer and Communications Security, ASIA CCS '13, Hangzhou, China*, 08–10 May 2013, pp. 329–334

Chapter 3
Fingerprinting Android Malware Packages

A fuzzy (hashing) or approximate fingerprint of binary software is a digest that captures its static content, in similar manner to cryptographic hashing fingerprints such as MD5 and SHA1. Still, the fuzzy fingerprint change is virtually linear to the change in the binary content. In other words, smaller changes in the static content of the malware will cause a minor change in the computed fuzzy fingerprint. In the context of cybersecurity, this is an important property that helps in detecting polymorphic malware attacks. Current fuzzy fingerprints such as *ssdeep*[1] are computed for the app binary as a whole, which makes them less effective for detecting malicious app variations. This problem gets complicated in the case of Android OS due to the apps packaging structure, which contains not only the actual compiled code but also other content such as media files. To overcome this limitation, we propose an effective and broad fuzzy fingerprint that captures not only binary features but also the underneath structure and semantics of the *APK* package.

Accordingly, our approach for computing Android app fingerprints, as shown in Fig. 3.1, relies on decomposing the actual *APK* file into different content categories. For each category, we compute a customized fuzzy hash (sub-fingerprint). Note that for some categories, for instance, *Dex* file, the application of the customized fuzzy hashing on the whole category content does not capture the structure of the underlying category. In this case, we apply fuzzy hashing against a selected *N-grams* of the category content. In our context, we use *byte n-grams* on binary files and *instruction n-grams* on assembly files. Furthermore, a best practice in malware fingerprinting is to increase the entropy of the app package content [1]. To this end, we compress each category, as proposed in [1] to increase the entropy, content before computing the customized fuzzy hash on its N-grams. The resulting fuzzy hashes (sub-fingerprints) are then concatenated to produce the final composed fuzzy fingerprint, called **APK-DNA**. As depicted in Fig. 3.1, there are two main processes: First, we build a database of fingerprints by generating **APK-DNA**s of

---

[1] https://ssdeep-project.github.io/ssdeep/index.html.

© The Author(s), under exclusive license to Springer Nature Switzerland AG 2021
E. B. Karbab et al., *Android Malware Detection Using Machine Learning*, Advances in Information Security 86, https://doi.org/10.1007/978-3-030-74664-3_3

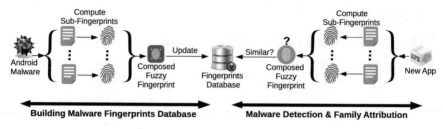

Fig. 3.1 Approximate fingerprint approach

known malware samples to identify whether a new app is malicious or not and to attribute the family of malicious ones. Second, for each malware under investigation, we compute its **APK-DNA** and match it against existing fingerprints in the database of known malware samples.

By generating fuzzy fingerprints for all known malware, in this scenario, we will able to fingerprint of known malware to make the detection. Thus, the detection process starts by computing **APK-DNA** fingerprints for known Android malware. We use multiple compression schemas for testing purposes. Thus, in the final fingerprint of the *APK*, only one compression is used. Moreover, we use **APK-DNA** fingerprints as a basis to design and implement **ROAR**, a novel framework for malware detection and family attribution. **ROAR**'s first approach, namely *family-fingerprinting*, computes a fingerprint for each malware family. Afterward, it uses these family fingerprints to make security detection decisions on new apps. In the second approach, *peer-matching*, **ROAR** uses the whole fingerprint database for detection and attribution.

3.1   Approximate Static Fingerprint

In this section, we present our approach for Android apps fingerprints generation.

3.1.1   Fingerprint Structure

We leverage the aforementioned *APK* structure to define the most important components for fingerprinting. The design of the *APK* fingerprint must consider most of its important components as unique features to distinguish between different malware samples. As depicted in Fig. 3.2, **APK-DNA** is composed of three main sub-fingerprints based on their content type: **Metadata**, **Binary**, and **Assembly**. The **Metadata** sub-fingerprint contains information, which is extracted from the **AndroidManifest.xml** file. We particularly focus on the required permissions. The aim is to fingerprint the permission used for a specific Android malware or

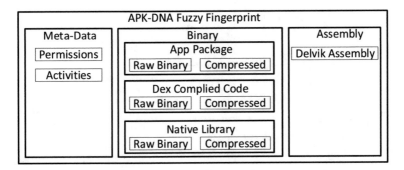

Fig. 3.2 Android package fingerprint structure

malware family. Our intuition stems from the fact that some Android malware samples need specific types of permissions to conduct their malicious actions. For example, the **DroidKungFu1** malware [2] requires access to personal data to steal sensitive information. Having Android malware without access permissions to personal data, e.g., *phone number*, would suggest that this malware is most likely not part of **DroidKungFu1** family. Other metadata information could be considered for malware segregation, for instance, *Activity* list, *Service* list, and *Component*. In the current design of APK-DNA, we focus on the required access permissions.

The **Binary** sub-fingerprint captures the binary representation of the *APK* file content. In other words, we aim to fingerprint the byte sequence of Android malware. In this context, we use *n-grams* [1] as we will present in Sect. 3.1.2. We divide the binary sub-fingerprint into three parts: **App Package**, **Dex Compiled Code**, and **Native Library**. The **App Package** consists of the *APK* file. Thus, all the components inside the package are considered (e.g., media file). Along with the raw *APK* package, we apply a compression schema on the package to increase its *entropy* [1]. In the **Dex Compiled Code**, we focus on the code section of the Android malware, which is located in the *Dex* file of Android apps. The use of the code section for malware detection has proven its accuracy [3]. In the context of Android malware, we use extracted features from the *classes.dex* as part of the APK-DNA. Besides, by applying compression, we use a high-entropy version of the *classes.dex* for fingerprinting. The **Native Library** part of the binary sub-fingerprint captures C/C++ shared libraries, used by malware. Using the native library for malware fingerprinting is essential in some cases, for example, to distinguish between two Android malware samples. For instance, if the malware uses a native library, it is more likely to be **DroidKungFu2** rather than **DroidKungFu1** because **DroidKungFu2** malware family uses C/C++ library and **DroidKungFu1** uses only *Java bytecode*.

In the **Assembly** sub-fingerprint, we also focus on the code section of Android malware, which is *classes.dex*. However, we do not consider the binary format. Instead, we use the reverse engineer assembly code. As we will present in Sect. 4.4.1, we reverse engineered the *Dalvik bytecode* in order to extract instruction

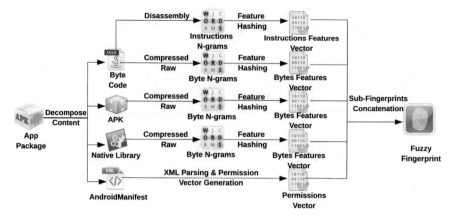

Fig. 3.3 Fingerprints computation process

sequences used in the app. The **Assembly** sub-fingerprint aims to discriminate malware using the unique instruction sequences in the assembly file. We use the same technique as in the **Binary** sub-fingerprint, i.e., *n-grams*. However, here we consider the assembly instructions instead of bytes. In addition to assembly instructions, we could also consider *section names*, *call graphs*, etc. In the current design, we focus on the assembly instructions for fingerprinting.

3.1.2   Fingerprints Generation

In this section, we present the steps required to generate APK-DNA fingerprints, as shown in Fig. 3.3. In addition, we present the main techniques adopted in the design of the fingerprint, namely *N-gram* and *Feature Hashing*. Afterward, we show the similarity techniques that are employed to compare *APK* fingerprints.

3.1.2.1   N-grams

The *N-gram* technique is used for computing contiguous sequences of N items from a large sequence. For our purpose, we use N-grams to extract the sequences (the order is important within the n-gram sequence) used by Android malware to be able to distinguish between different malware samples. To increase the fingerprint accuracy, we leverage two types of N-grams, namely *instruction N-grams* and *bytes N-grams*. As depicted in Fig. 3.4, the instruction N-grams are the unique sequences in the disassembly of a *Dex* file, where instructions are stripped from the parameters. In addition to instruction N-grams, we also use byte N-grams on different contents of the Android package. Figure 3.4 illustrates different N-grams on both instructions and bytes of the first portion of the **AnserverBot** malware. We have experimented

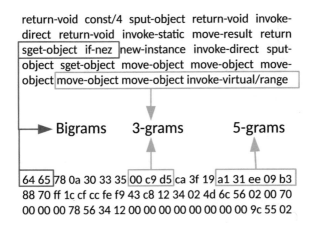

Fig. 3.4 Instructions and bytes from AnserverBot malware

with multiple options such as *bigrams*, *3-grams*, and *5-grams*. The last one provided the best results in the design of **APK-DNA** fingerprint, as will be shown in the evaluation section. The result of N-grams extraction is the list of unique *5-grams* for each content category, i.e., *assembly instructions*, *classes.dex*, *native library*, and *APK file*.

3.1.2.2   Feature Hashing

Feature hashing is a machine learning preprocessing technique for squashing an arbitrary number of features into a fixed-size feature vector. The feature hashing algorithm, described in Algorithm 1, takes as input the set of sequences generated by applying the N-gram technique and the length of the output feature vector. In the current implementation of **APK-DNA**, we use a bit feature vector of 16KB. However, the size could be adjusted according to the needed density of the bit feature vector to distinguish between apps. For example, the size of the assembly instruction vector could be less than the dex vector since the density produced by the instruction content is less than the dex one. Notice that in our implementation, we store only a binary value, which defines whether the N-gram exists or not. The standard feature hashing uses the frequency, i.e., the number of occurrences of a given N-gram. The output of the feature hashing algorithm is a feature-bit vector. Instead of using existing fuzzy hashing algorithms such as *ssdeep*, we leverage the feature vector as our fuzzy hashing technique for implementing **APK-DNA** fingerprint. In the next section, we present the complete process of computing the fingerprint using N-grams and feature hashing as basic blocks.

Algorithm 1: Feature vector computation

---

input : **N-grams**: Set,
         L: Feature Vector Length
output: Binary Feature Vector
features_vector = **new** bitvector[**L**];
for *Item* *in N-grams* **do**
  | H = hash(**Item**) ;
  | feature_index = H mod **L** ;
  | features_vector[feature_index] = 1 ;
end

3.1.2.3   Fingerprint Computation Process

As shown in Fig. 3.3, the fingerprint computation process starts by decomposing the Android app *APK* file into four different content categories: (1) *Dalvik bytecode*, (2) *APK file*, (3) *native libraries*, and (4) *AndroidManifest file*. Each binary content is compressed to increase the entropy. Afterward, we extract the byte N-grams from the raw assembly and the compressed content. The resulting set of N-grams is provided as input to the feature hashing function to produce the customized fuzzy hashing. The size of each customized fuzzy hash is 16KB, as mentioned in Sect. 3.1.2.2. For *Dalvik bytecode*, we fingerprint the assembly code in addition to the binary fingerprint. First, we reverse engineer the *Dex* file to produce its assembly code. After preprocessing the assembly, we use the instruction sequence of the Android app to extract the instruction N-grams set. Afterward, we use feature hashing to generate a 16KB bit vector fingerprint for the assembly code. The current design of **APK-DNA** uses the *feature hashing* technique without *feature selection* because we aim to keep maximum information on the targeted malware instance or its family. However, *feature selection* could be a promising technique to explore in future **APK-DNA** design.

Regarding the *AndroidManifest file*, we first convert it into a readable format, then parse it to extract the required permissions by the Android app. To use the required permission app fingerprinting, we use a bit vector of all Android permissions in a predefined order. For each given required permission, we flag the bit to 1 in the permission vector if it exists in the *AndroidManifest file*. The result is a bit vector for all the permissions of the Android app. At the end of the operations mentioned above, we generate five-bit vectors. The final step of the fuzzy fingerprint computation consists of concatenating all the produced digests into one fingerprint, designated as **APK-DNA**. It is important to mention that, for similarity computation, we also keep track of the bits of each content vector. Notice that the content categories are mandatory for Android apps except the *native library*, which may not be part of the app. Therefore, we use a bit vector of zeros for the feature vector of the *native library*. The final size of **APK-DNA** is 16KB for the feature vector of each content (there are four feature vectors: assembly, bytecode, APK, and native library). However, for the permission vectors, we use a 256-bit feature vector since the Android permission system does not exceed this number.

3.1.2.4 Compute Fingerprints Similarity

The main reason for adopting the feature vector as a customized fuzzy hash is to make the similarity computation straightforward using *Jaccard Similarity*, as shown in Eq. (3.1). Since we have a set of bit feature vectors flagging the existence of a feature, we adopt a *bitwise Jaccard* similarity, as depicted in Eq. (3.2). The *Jaccard Similarity* is computed by dividing the cardinality of the intersection set by the cardinality of the union set.

$$Jaccard(X, Y) = \frac{|X \bigcap Y|}{|X \bigcup Y|}$$
$$0 \leq Jaccard(X, Y) \leq 1 \tag{3.1}$$

$$Jaccard\_bitwise(A, B) = \frac{Ones(A \wedge B)}{Ones(A \vee B)}$$
$$0 \leq Jaccard\_bitwise(X, Y) \leq 1 \tag{3.2}$$

Let A and B be two bit feature vectors; the union of the two vectors is given by the logical expression $A \vee B$, and its cardinality is the number of "1" bits in the resulting vector. Similarly, the cardinality of the intersection of the two vectors is the number of "1" bits in $A \wedge B$ bit vector. As presented in Sect. 3.1.2.3, APK-DNA fuzzy fingerprint is composed of five fuzzy hashes, which are bit feature vectors. To compute the similarity between two fingerprints, we calculate the bitwise Jaccard similarity between the bit feature vectors representing the same content. In other words, we calculate the similarity between the feature vectors of the assembly, bytecode, APK, native library, and permissions. The result is a set of five similarity values.

3.2 Malware Detection Framework

In this section, we leverage the proposed APK-DNA fingerprint for Android malware detection. More precisely, we present (1) the *family-fingerprinting* approach, where we define and use a family-fingerprint, and (2) the *peers-matching* approach, where we compute the similarity between malware fingerprints. Both approaches are based on the *peer-fingerprint voting* mechanism to decide on malware detection and family attribution.

3.2.1 Peer-Fingerprint Voting

As we have seen in Sect. 3.1.2.4, comparing two Android malware packages consists of computing similarities between their metadata, binary, and assembly sub-fingerprints, which gives numerical values on how the two packages are similar in a specific content category, as presented in Algorithm 2. In addition, we add the summation of all the similarities as a summary value of these sub-contents similarities. Note that other summary values, such as the average and the maximum, could also be used. However, it is challenging to detect the most similar packages if we compare an unknown package to known malware packages using multiple sub-fingerprints. The most obvious solution is to merge bit vectors of each content category into one vector and then compute the similarity of the resulting feature vector. However, in our case, merging bit vectors will heavily reduce the contribution of some sub-fingerprints in the similarity computation.

Algorithm 2: APK-DNA similarity computation

---
input : APK-DNA **A**: list
        APK-DNA **B**: list
output: similarity-list: list
similarity-list = *empty-list( )*;
for *content in content categories* **do**
    similarity = Jaccard_bitwise(A[content],B[content]) ;
    similarity-list.add(similarity);
end
summation = sum(similarity-list);
similarity-list.add(summation);

Algorithm 3: Peer-fingerprint voting mechanism

---
input : similarity-list **A-B**: list
        similarity-list **A-C**: list
output: Decision
A-B-count = 0 ;
A-C-count = 0 ;
for *content in content categories* **do**
    if *A-B[content] >A-C[content]* **then**
        A-B-count += 1;
    else
        A-C-count += 1;
    end
end
if *A-B-count >A-C-count* **then**
    Decision = A-B;
else
    Decision = A-C;
end

Fig. 3.5 Peer-fingerprint
voting

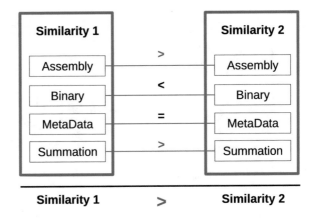

Likewise, the density of the assembly feature vector is considerably less compared to the binary feature vector. Consequently, we propose to use a composed similarity using *peer-fingerprint voting*. The idea is to compare parts (sub-fingerprints) instead of comparing full fingerprints, as depicted in Algorithm 3. In other words, we examine each sub-similarity pairs. The decision is made by a voting mechanism on the result of each sub-comparison. Moreover, in case of equal votes, we compare the *summation* of the sub-similarities to remove the ambiguity as shown in the example depicted in Fig. 3.5. At this stage, we can compare different Android packages and decide on the most similar package to a given one. In what follows, we propose two approaches to malware detection.

3.2.2   Peer-Matching

In the *peer-matching* approach, ROAR queries the fingerprints database to check the most similar malware fingerprint. To detect Android malware variation, we build a *malware fingerprint* database by computing APK-DNA for known Android malware. The more fuzzy fingerprints in this database, the broader the detection system could cover. As shown in Fig. 3.6, for each new malware, we compute its APK-DNA and add it to the database.

To attribute the malware family to a new app, we first compute the similarity between the malware fingerprint and each entry in the database of known malware fingerprints, as depicted in Fig. 3.6. To this end, we use *bitwise Jaccard* similarity, presented in Sect. 3.1.2.4, to produce a set of sub-similarity values, i.e., the *composed similarity*. Afterwards, to compare the *composed similarity* values, we use the previously presented *peer-voting* technique. The entry with the highest similarity value that exceeds an acceptance threshold determines the malware family. In the current implementation, we use an experimentally derived static threshold. As such, *Peer-matching* is a simple approach for malware detection and family attribution.

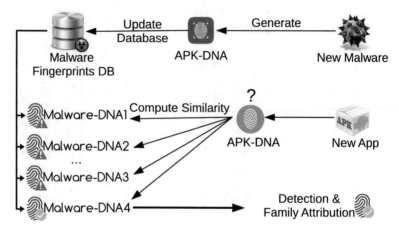

Fig. 3.6 Malware detection using peer-matching

3.2.2.1 Family-Fingerprinting

In this approach, some extra steps are needed to build a second database of malware family fingerprints. The aim is to reduce the number of database entries required to match an Android malware fingerprint. For this reason, we propose a custom approximate fingerprint for a malware family. The intent is to leverage this family-fingerprint for malware detection purposes. The idea is to build a database of family fingerprints from known Android malware samples, and use this database for similarity computation with unknown malware apps. The number of malware families limits the actual size of the family-fingerprint database. Notice that the fingerprint structure for a malware family is the same as for a single malware, i.e., metadata, binary, and assembly family sub-fingerprints.

Algorithm 4: Family-fingerprint computation

---

input : **Malware Family X Fingerprints**: Set
output: **Family X Fingerprint**: **FP_X**
FP_X = **new** bitvector[Zeros];
for *fprint in Fingerprints* **do**
    FP_X{meta} = **FP_X**{meta} **or fprint**{meta};
    FP_X{bin} = **FP_X**{bin} **or fprint**{bin};
    FP_X{asm} = **FP_X**{asm} **or fprint**{asm};
end

Algorithm 4 depicts the computation of the family-fingerprint based on the underlying content sub-fingerprints. First, the fingerprint is initialized to zeros (each content sub-fingerprint). Afterward, the fingerprint is generated by applying a logical *OR* on the current value of the family-fingerprint with a single malware

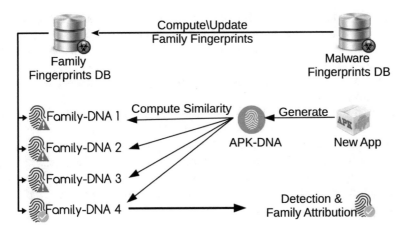

Fig. 3.7 Malware detection using family-fingerprint

fingerprint. Note that each content sub-fingerprint is computed separately. This operation is applied to all malware samples in the database. After calculating the fingerprints from known malware samples, we store them in a *family-fingerprint* database, which is used for detection and family attribution. The detection process is composed of several steps. First, for a given Android package, we generate its fingerprint as described in Sect. 3.1.2.4. Then, we compute the similarity between this fingerprint and each family-fingerprint in the database. The family with the highest similarity score will be chosen as the family of the new app if the similarity value is above a defined threshold. In the current implementation, we use an experimentally derived static threshold, which is only applied to the *summation* part of the composed similarity. The result is similar to the single malware fingerprint, but it represents a malware family instead of a particular malware as depicted in Fig. 3.7.

3.3 Experimental Results

In this section, we present the results in terms of accuracy for both approaches that are adopted in **ROAR** framework, namely family-fingerprinting and peer-matching.

3.3.1 Testing Setup

Our dataset contains 928 malware samples from Android Malware Genome Project [2, 4]. For evaluation, we selected malware families with many samples since some malware families in Android Malware Genome Project [4] contain only a few samples (some families have only one sample), as depicted in Table 3.1. Clearly, by filtering out other families that do not have enough malware samples, we may miss the detection of these malware families. In addition to known malware samples, we use benign Android applications in each evaluation. These apps have been downloaded from Google play randomly without considering the popularity of the app, as shown in Table 3.1.

For each evaluation benchmark and from the balanced dataset, we randomly sample 0% (70 for training and 30% for testing is a common split method in machine learning) of each family from the dataset to build the fingerprints database. The rest of the dataset (30%) is used for the evaluation of ROAR approaches and sub-fingerprints. Notice that the random sampling is done for every benchmark evaluation. Accordingly, we repeat the assessment five times. The final result is the average of the evaluation results.

3.3.2 Evaluation Results

In this section, we present the evaluation results of the ROAR framework. Each approach is separately evaluated. The results are presented using *F1-score*, *precision*, and *recall*. The approach is evaluated multiple times (five) using different fingerprint setups, i.e., combinations of sub-fingerprints, which are used to compute the similarities using *peer-voting* technique. Furthermore, we present a comparison between the proposed *peer-voting* similarity technique and the merged fingerprint similarity.

Confusion Matrix Description In addition to the previous evaluation metrics, we also use the confusion matrices in each evaluation, as shown in Figs. 3.8 and 3.9. Each confusion matrix is a square table, where the number of rows and columns are, respectively, malware families and benign apps following the same order as in Table 3.1. The columns and rows from 0 to 7 are, respectively, the malware families, **AnserverBot, KMin, DroidKungFu4, GoldDream, Geinimi, BaseBridge, DroidDreamLight**, and **DroidKungFu3**, and the column and row

Table 3.1 Evaluation malware dataset

#	Malware family/apps	Number of samples
0	AnserverBot	187
1	KMin	52
2	DroidKungFu4	96
3	GoldDream	47
4	Geinimi	69
5	BaseBridge	122
6	DroidDreamLight	46
7	DroidKungFu3	309
8	Benign apps	100

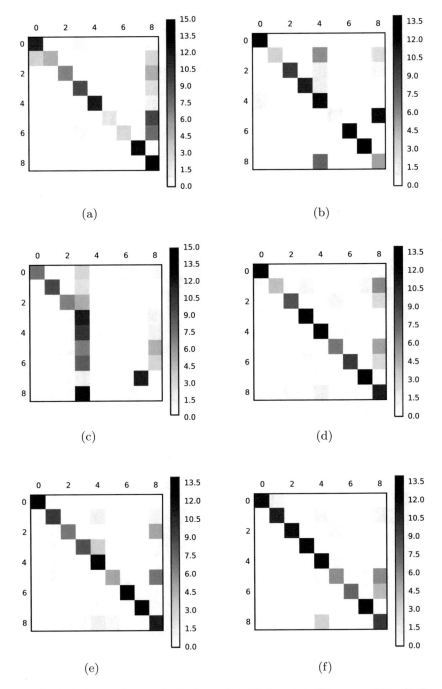

Fig. 3.8 Confusion matrices of family-fingerprint. (**a**) Assembly (F1-score = 69%). (**b**) Permission (F1-score = 69%). (**c**) Dex (F1-score = 41%). (**d**) Assembly, permission, Dex, APK (F1-score = 81%). (**e**) Assembly, permission (F1-score = 82%). (**f**) Assembly, permission, Dex (F1-score = 85%)

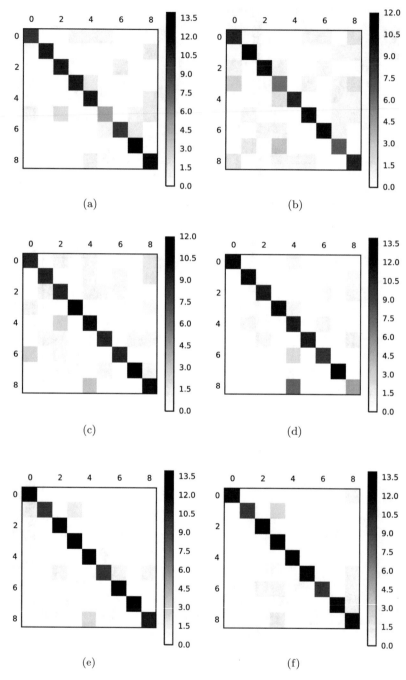

Fig. 3.9 Confusion matrices of peer-matching approach. (**a**) Assembly (F1-score = 91%). (**b**) Permission (F1-score = 81%). (**c**) Dex (F1-score = 80%). (**d**) Assembly, permission, Dex, APK (F1-score = 84%). (**e**) Assembly, permission (F1-score = 94%). (**f**) Assembly, permission, Dex (F1-score = 93%)

8 represent the benign apps. The interpretation of the confusion matrix results is related to the intensity of the color in its *diagonal*. The darker the color is in the *diagonal*, the higher and the more accurate are the results of the evaluation (*true positive*). The color intensity of the confusion matrix cells represents the number of malware/apps that have been assigned to this cell. However, the less intense is the color in the *diagonal*, and the more intense in the other cells, the less accurate is the result.

False Negative For any row from 0 to 7 (i.e., malware family), there is a missing malware family attribution if we have a gray color in the other cells of the same row. Even though we missed in the family attribution task, we still detect the app as malicious. However, a gray cell in column 8 (benign apps) means that we missed both detection and family attribution (*detection false negative*). **False Positive**. In the benign apps row (row number 8), the gray color in malware cells indicates that there is a *false positive*. In other words, there are benign apps that are detected as malicious. The number of *false positives* could be measured using the intensity of the color according to the color bar.

3.3.2.1 Family-Fingerprinting Results

As depicted in Table 3.2, the F1-score, precision, and recall of the *family-fingerprinting* vary according to the fingerprint setup. We evaluated the approach for each content type separately, i.e., *assembly*, *permission*, and *dex* files, so that we can clearly see the impact of each component in the final fingerprint. Both *assembly* and *permission* types show more accurate results compared to *dex* type. Specifically, the *permission* shows a promising result (82% precision), as illustrated in Table 3.2. This indicates the impact of the metadata on Android malware detection. It is that investigating other metadata could result in higher accuracy. The *APK* fingerprint results are surprising because of the poor accuracy value, under 40% f1-score. The learned lesson is that applying the fuzzy fingerprinting (including *ssdeep*) to the

Table 3.2 Accuracy results of the family-fingerprinting approach

Fingerprint setup	F1-score	Precision	Recall
Assembly	76%	88%	68%
APK	33%	36%	32%
Permission	76%	84%	70%
Dex	44%	46%	43%
Assembly, permission, Dex, APK	83%	88%	80%
Assembly, permission	84%	88%	81%
Assembly, permission, Dex	**86%**	**89%**	**84%**
Best fingerprint setup	**86%**	**89%**	**84%**

The bold values represent the best values

whole package could mislead the malware investigation when using fuzzy matching. The confusion matrix for each setup demonstrates a more granular view of the result, as shown in Fig. 3.8, where the indexes are the malware families (Table 3.1). On the other hand, the combination setups indicate accurate results compared with single content fingerprints. We depict three sub-fingerprints, which correspond to the best results. Note that the setup composed of *assembly*, *permission*, and *dex* bytecode shows the highest *F1-Score*.

3.3.2.2 Peer-Matching Results

Peer-matching shows a higher F1-score, precision, and recall for all the setups compared to *family-fingerprinting*, as shown in Table 3.3. This can be clearly seen in the confusion matrices in Fig. 3.9. In contrast to the previous results, the *dex* bytecode shows a higher *precision* than *assembly* and *permission*, but it is still lower in both *recall* and *F1-score*. The setup combination (*assembly*, *permission*) has the highest accuracy in the *peer-matching* approach. As such, using only two content categories, metadata *permission* and *assembly* instruction sequences, we achieve a very promising detection rate, especially considering that the computation of these sub-fingerprints is light and simple compared to the state-of-the-art fingerprint hashing techniques.

3.3.2.3 Peer-Voting vs Merged Fingerprints

As presented in Sect. 3.2.1, the most obvious technique to deal with multiple sub-fingerprints is to merge all of them (*merged fingerprint*). However, we propose *peer-voting* to compare multiple sub-fingerprints and use the majority voting to confirm the similarity. To test the proposed technique, we evaluate it against the *merged fingerprint* for the same fingerprinting setup. As shown in Table 3.4, *peer-voting* shows a higher accuracy than the merging one. A more illustrative view of the result can be seen in the confusion matrix in Fig. 3.10.

Table 3.3 Accuracy result of peer-matching approach

Fingerprint setup	F1-score	Precision	Recall
Assembly	91%	91%	90%
Apk	46%	48%	44%
Permission	81%	82%	80%
Dex	86%	90%	84%
Assembly, permission, Dex, APK	85%	91%	81%
Assembly, permission, Dex	93%	94%	93%
Assembly, permission	**94%**	**95%**	**94%**
Best fingerprint setup	**94%**	**95%**	**94%**

The bold values represent the best values

Table 3.4 Accuracy result using merged fingerprint

Fingerprint setup	F1-score	Precision	Recall
Merged in family-approach	77%	84%	72%
Peer-voting in family-approach	**85%**	**89%**	**84%**
Merged in peer-approach	87%	87%	86%
Peer-voting in peer-approach	**94%**	**95%**	**94%**

The bold values represent the values of Peer-voting in family-approach that are the best compared to other techniques

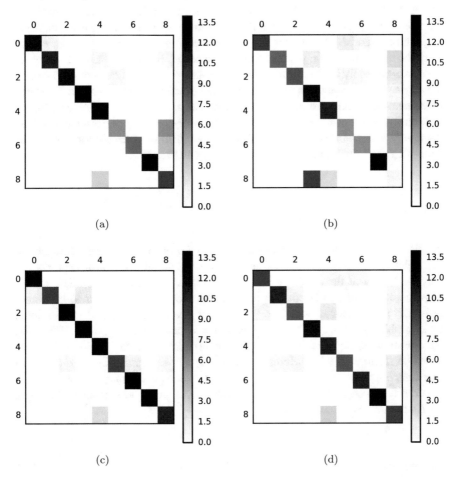

Fig. 3.10 Confusion matrices and F1-score using merged fingerprint and peer-voting. (**a**) Voting in family (85%). (**b**) Merged in family (72%). (**c**) Voting in peer-matching (94%). (**d**) Merged in peer-matching (86%)

3.3.3 Discussion

Similarity computation based on *family-fingerprinting* is bounded by the number of families, which is a way less than the number of malware samples. Consequently, this approach is more scalable compared to *peer-matching*. However, *family-fingerprinting* suffers from two main drawbacks. First, its design is more complicated compared to *peer-matching* since it requires an extra step to build family-fingerprint database, even though we do not create the database for every new matching. Second, because of the rapid change of the malware database, we continuously need to update the family-fingerprint database with the latest malware fingerprint features. On the contrary, *peer-matching* does not require an update since the new malware fingerprint is directly inserted. Regarding the *peer-matching* approach, it mainly suffers from scalability issues since, for each new malware, all the entries in the database need to be matched. Therefore, query latency is related to the number of malware fingerprints in the database which we will address in the next chapter. However, in terms of accuracy, *peer-matching* performs better than the *family-fingerprinting* one. Therefore, the application of each approach can be defined by the actual trade-off between latency and accuracy.

3.4 Summary

In this chapter, we have presented a thorough fuzzy fingerprinting approach for investigating Android malware variations. The proposed fingerprint captures not only the binary of the *APK* file but also both its structure and semantics. This is of paramount importance since it allowed the proposed fingerprint to be highly resistant to app changes, which is a significant advantage compared to traditional fuzzy hashing techniques. Moreover, we have leveraged the proposed APK-DNA fingerprint to design and implement a novel framework for Android malware detection, following two different approaches, namely *family-fingerprinting* and *peer-matching*. In the next chapter, we have leveraged the proposed APK-DNA fingerprint to design and implement an innovative, efficient, and scalable framework for Android malware clustering.

References

1. M.M. Masud, L. Khan, B.M. Thuraisingham, A scalable multi-level feature extraction technique to detect malicious executables. Inf. Syst. Front. **10**(1), 33–45 (2008)
2. Y. Zhou, X. Jiang, Dissecting android malware: characterization and evolution, in *IEEE Symposium on Security and Privacy, SP 2012, San Francisco, California, USA*, 21–23 May 2012, pp. 95–109

3. J. Jang, D. Brumley, S. Venkataraman, Bitshred: feature hashing malware for scalable triage and semantic analysis, in *Proceedings of the 18th ACM Conference on Computer and Communications Security, CCS 2011, Chicago, Illinois, USA*, 17–21 Oct 2011, pp. 309–320
4. Android Malware Genome Project, http://www.malgenomeproject.org/. Accessed Jan 2015

Chapter 4
Robust Android Malicious Community Fingerprinting

Security practitioners can combat large-scale Android malware by decreasing the *analysis window* size of newly detected malware. The window starts from the first detection until signature generation by anti-malware vendors. The larger the window is, the more time the malicious apps are given to spread over the users' devices. Current state-of-the-art techniques have a large analysis window due to the significant number of Android malware appearing daily. Besides, these techniques use manual analysis in some cases to investigate malware. Therefore, decreasing the need for manual detection could significantly reduce the *analysis window*. To address the aforementioned issue, we elaborate systematic techniques and tools for the detection of both known family apps and new malware family apps (i.e., variants of existing families or unseen malware). To do so, we rely on the assumption that any pair of Android apps, with distinct authors and certificates, are most likely to be malicious if they are highly similar. Because the adversary usually repackages multiple app packages with the same malicious payload to hide it from anti-malware and vetting systems. Consequently, it is difficult to detect such malicious payloads from benign functionalities of a given Android package. Accordingly, a pair of Android apps should not be very similar in their components, excluding popular libraries. This observation, as mentioned earlier, could be used to design and develop a security framework to detect Android malware apps .

In this chapter, we propose a novel Android app fingerprinting technique, APK-DNA, inspired by fuzzy hashing. We specifically target fingerprinting Android malicious apps. Computing the APK-DNA of a suspicious app requires a low computation time. Afterward, we leverage the previously mentioned assumption (i.e., very similar apps might be malware from the same malware family) to propose a cyber-security framework, namely Cypider (*Cyber-Spider for Android malware detection*), to detect and cluster Android malware without prior-knowledge of Android malware apps . Cypider consists of a novel combination of a set of techniques to address the problem of Android malware, clustering, and fingerprinting. First, Cypider can detect repackaged malware (malware families), which constitute

E. B. Karbab et al., *Android Malware Detection Using Machine Learning*, Advances in Information Security 86, https://doi.org/10.1007/978-3-030-74664-3_4

the vast majority of Android malware apps. Second, it can detect new malware apps, and more importantly, Cypider performs the detection automatically and in an unsupervised way (i.e., no prior-knowledge about the apps). The fundamental idea of Cypider relies on building a *similarity network* between the targeted apps static content in terms of fuzzy fingerprints. Actually, Cypider extracts, from this *similarity network*, sub-graphs with high connectivity, called *communities*, which are most likely to be *malicious communities*.

4.1 Threat Model

In the context of this chapter, the focus is on detecting malware targeting Android mobile apps with no prior-knowledge about the malware. In particular, instead of focusing on the detection of an individual instance of malware, Cypider targets bulk detection of malware families and variants as malicious communities in the similarity network of the apps dataset. Moreover, Cypider aims for a scalable yet accurate solution that can handle the overwhelming volume of the daily detected malware, which could aggressively exploit users' smart devices. Cypider is robust (Sect. 4.11) but not immune against obfuscated apps contents. Cypider could handle some types of obfuscations because it considers different static contents of the Android package in the analysis. This makes Cypider more resilient to obfuscation as it can group malware apps by also considering other static contents that are not obfuscated, such as app permissions or Android API calls.

4.2 Usage Scenarios

In the context of this book, Cypider has two primary usage scenarios. In the first scenario, Cypider can be applied only to malicious Android apps. The aim is to speed up the analysis process and attribute malware to their corresponding families. Under the first scenario, the overall malware analysis process is boosted by automatically identifying malware families and minimizing the overall manual analysis effort. The outcome of the previous process consists in the communities of malicious apps. The attribution of a family to a given community can be achieved by attributing a small set (one app in most cases) among its malicious apps. In the second scenario, Cypider is applied to mixed Android apps (i.e., malicious or benign). Such a dataset could be the result of a preliminary suspiciousness app filtering. Therefore, a lot of *false positives* can be recorded; we assume that benign apps—meaning *false positives*—constitute 50–75% of the actual suspicious apps. Based on the previous assumption, we could identify malicious Android apps by detecting and extracting app communities that share a common payload. We could infer that apps with high similarity are most likely to be malicious.

4.3 Clustering Process

Cypider framework uses a dataset of unlabeled apps (malicious or mixed) in order to produce *community fingerprints* for the identified *app communities*. Cypider overall process is achieved by performing the following steps, as illustrated in Fig. 4.1:

At the beginning of the Cypider process, we need to filter out apps developed by the same author; we call them *sibling apps*. We aim here to remove the noise of having app communities of sibling apps because they tend to have many similar features since authors reuse components across different apps. Cypider identifies sibling apps in the dataset based on their version, app hash, and author cryptographic signature (provided in the META-INF directory in the APK file). Therefore, we only keep apps with no duplication in the author identities since adversaries favor the use of multiple fake author identities to prevent the removal of all apps in case of detected maliciousness in one of them. Regarding numerous apps with the same author, Cypider randomly selects one app. Afterward, if the chosen app is recognized as malicious in the analysis results, Cypider will tag all its sibling apps as malicious.

After filtering the sibling apps, we need to derive from the actual app's package meaningful information that could identify the app and help to compute the similarity against other apps. For this purpose, Cypider extracts static features from the apps, which could be either benign or malicious, depending on the usage scenario. Feature engineering is the most critical part of the whole framework in the context of Android malware detection (other usage scenarios could have different static features, but the overall approach is the same). It is essential to mention that the selected features must be resilient to the attacker's deceiving techniques. To this

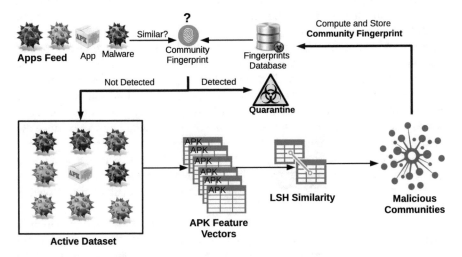

Fig. 4.1 Cypider framework overview

end, the features need to be broad enough to cover most of the static characteristics of a given Android APK. The more extensive the features are, the more resilient they are. For our purposes, we leverage static analysis features of the APK in the design of Cypider. In particular, we extract such features from each content category (classes.dex, resources, assembly, etc.), as described in Sect. 4.4.

Relying on the extracted features from each content, Cypider computes a fixed-length *feature vector* for each content features. In order to reduce and normalize the size of the feature vectors, we equip Cypider with a machine learning preprocessing technique, namely *feature hashing* [1] (or *hashing trick*), as presented in Sect. 4.4.7. As a result, Cypider produces, from the extracted features of the previous stage, multiple feature vectors with a small and fixed size. The number of the generated feature vectors depends on how many APK contents are used in feature extraction (each content type corresponds to one feature vector).

For efficient comparison between apps, we empower Cypider system with a highly scalable similarity computation system based on locality sensitive hashing (LSH) [2], which computes the similarities between apps, as presented in Sect. 4.5. Given a pair of apps, we calculate the similarity between each content *feature vector* from the previous stage to decide if they are connected or not from the perspective of that content. The result of this step is an undirected network (or similarity network), where the nodes are Android apps, and the edges represent the high similarity to one content between apps. For similar apps, multiple connecting edges are expected. Besides, the more the edges are, the more the apps are suspected to be malicious.

Cypider leverages a *similarity network* in order to detect malicious app communities. For malicious app, Cypider extracts highly connected app communities and then excludes these apps from the dataset. The remaining apps (i.e., apps that are not part of any community) are considered in another Cypider malware detection iteration. We expect to get a pure (only malware apps from the same family) or near-pure community if the containing apps of a given community have, respectively, the same or almost the same Android malware family. In the case of a mixed dataset, Cypider first excludes all the app nodes with a degree 1 (i.e., the app is only self-similar), which are most likely to be benign apps (or might be a zero-day threat). Afterward, Cypider extracts the apps of malicious communities.

The rest of apps will be considered in another Cypider iteration. At this point, we expect to have some benign communities as *false positives*. However, the similarity network makes Cypider's decision explainable (the security practitioner can check which static contents are similar between the detection communities) because the security practitioner can track which content these apps are similar to. The previous option could also help sharpening the static features to prevent benign apps from being detected in malicious communities. For community detection (Sect. 4.6), we adopt a highly scalable algorithm [3] to enhance Cypider's community detection module.

To this end, we consider a set of malicious communities, each of which is most likely to be a malware family or a subfamily. Cypider leverages these malicious communities to generate the so-called *community fingerprint* (Sect. 4.7) that captures the app features of a given detected community. Instead of using traditional crypto or fuzzy hashing of only one malware instance, we leverage a model produced by a one-class classifier [4], which provides a better-compressed format of a given Android malware family. This model is used to decide whether new malware apps are part of this family or not. The results consist of multiple *community fingerprints*, each of which corresponds to a detected community. The generated fingerprints are stored in the *signature database* for later use.

To this end, **Cypider** is ready to start another detection iteration with a new dataset, including the rest of unassigned apps from the previous iteration. The same previous steps will be followed for the new iteration. However, at this point, Cypider first checks the *feature vectors* of the new apps against the known *malware community fingerprints* stored in the database. The matched apps to a community fingerprint are labeled as malicious without adding them to the *active dataset*. Undetected apps are added to the *active dataset* and are considered in the next iteration of the detection process.

We consider **Cypider** approach as a continuous process, in which we detect and extract communities from the *active dataset* that always gets new apps (malware only or mixed with benign) daily, in addition to the rest of apps from the previous iterations.

4.4   Static Features

In this section, we present static features of Android packaging (*APK*). In this chapter, we only extract features from each app *APK* file using static analysis, to generate feature vectors, which are then used to compute the similarity with other apps feature vectors. In other words, the feature vector set will be the input to the LSH similarity computation module used to build the similarity network. As previously mentioned, the features should be broad enough to cover most of the static content of the APK file. Features could be categorized according to the main APK content types to: (1) Binary features, which are related to bytecode (Dex file) of the Dalvik virtual machine considering the hex dump of the Dex file along with the actual file. (2) Assembly features, which are computed from the assembly of *classes.dex*. (3) Manifest features, extracted from the Manifest file, which is vital to Android apps since it provides essential information about the app to the Android OS. (4) *APK features*, which include all the remaining APK file content, such as *resources* and *assets*. In this section, we present the static features based on the adopted concept to extract them (e.g., N-gram).

4.4.1   N-grams

The *N-gram* technique is used to compute contiguous sequences of N items from a large sequence. For our purpose, we use N-grams to extract the sequences derived from Android malware content with the aim to discriminate different malware samples. The N-grams from various Android app package contents, such as *classes.dex*, reflect the APK patterns and implicitly capture the underlying Android package semantics. We compute multiple feature vectors for each APK content. Each vector $V \in D$ ($|D| = \Phi^N$ where Φ represents all the possibilities of a given APK content). Each element in the vector V contains the number of occurrences of a particular APK content N-gram.

4.4.1.1   Classes.dex Byte N-grams

To increase the extracted information, we leverage two types of N-grams, namely *opcode N-grams* and *byte N-grams*, which are extracted from the binary *classes.dex* file and its assembly, respectively. From the *hexdump* of the *classes.dex* file, we compute *Byte N-grams* by sliding a window of the hex string as depicted in Fig. 3.4.

4.4.1.2   Assembly Opcodes N-grams

The opcode N-grams are unique sequences in the disassembly of *classes.dex* file, where the instructions are stripped from their operands. We choose opcodes instead of full instructions for multiple reasons: (1) Using opcodes tends to be more resilient to simple obfuscations that modify some operands such as hard-coded IPs or URLs. (2) Opcodes could be more robust to modifications, caused by repackaging, that alter or rename some operands. (3) In addition to being resilient to changes, opcodes can be efficiently extracted from Android apps.

The gained information from opcode N-grams could be increased by considering only functions that use a sensitive APIs such as SMS API. Also, excluding the most common opcode sequence decreases the noise in N-gram information. Also, the number of N-grams has a significant influence on the gathered semantics. The result of N-gram extraction is the list of unique N-grams with the occurrence number for each content category, i.e., *opcode instructions*, *classes.dex*. In addition to the opcodes, we also consider the *class names* and the *methods' names* as assembly features.

4.4.2   Native Library N-grams

The *Native Library* is part of the binary sub-fingerprint, which captures C/C++ shared libraries [5] used by malware. Using the native library for malware fingerprinting is essential in some cases to distinguish between two Android malware samples. For instance, if the malware uses a native library, it is more likely to be **DroidKungFu2** than **DroidKungFu1** because **DroidKungFu2** malware family uses C/C++ library and **DroidKungFu1** uses only *Java bytecode*.

4.4.2.1 APK N-grams

The N-gram of the APK file can give an overview of the APK file semantics. For instance, most of the repackaged apps are built from an original app with minor modifications [6]. Consequently, applying N-gram analysis on the APK file can detect a high similarity between the repackaged app and the original one. Besides, some components of the APK file, e.g., images and GUI layout structures, are preserved by the adversaries, especially if the purpose of the repackaging process is to develop a phishing malware. Both apps, in this case, are visually similar, and hence the N-gram sequences computed from both apps will be similar in the zone related to the resource directory.

4.4.3 Manifest File Features

In our context, *AndroidManifest.xml* is a source of essential information that could help in identifying malicious apps. The *permissions* required by apps are the most important features. For example, apps that require *SMS send* permission are more suspicious than other apps since a big portion of Android malware apps target sending SMS to premium charging phone numbers. In addition, we extract other features from *AndroidManifest.xml*, namely *activities*, *services*, and *receivers*.

4.4.4 Android API Calls

The required permissions provide a global view of possible app behaviors. However, we could get a more granular view by tracking *Android API calls*, knowing that one permission could allow access to multiple *API calls*. Therefore, we consider the *API* list used by the apps as the feature list. Furthermore, we use a filter list of *API* of the *suspicious APIs*, such as *sendTextMessage()* and *orphan APIs*, which are part of an undeclared permission. On the other hand, we extract the list of permissions, where none of their *APIs* has been used in the app.

4.4.5 Resources

In this category, we extract features related to *APK* resources, such as text strings, file names, and their content. An important criterion when filtering the files is to exclude the names of standard files, e.g., *String.xml*. Also, we include files' contents by computing **MD4** hashes on each resource file. At first glance, it seems that the use of MD4 is not convenient compared to more modern cryptographic hashing algorithms such as MD5 and SHA1. However, we choose the MD4 purposely because it is cheap in terms of computation. This allows to enhance the scalability of the system, yet, we achieve the goal of the file comparison between the malicious apps of the active dataset. Finally, we make a text string selection in the text resources, where we leverage *tf-idf* (term frequency-inverse document frequency) [7] technique for this purpose.

4.4.6   APK Content Types

Table 4.1 summarizes the proposed feature categories based on APK contents. It also depicts the features considered in the current implementation of **Cypider**. We believe that the used features give a more accurate representation of Android packages as we showed in Sect. 3.1. On the other hand, the features that we excluded such as *Text Strings* and *Assembly Class Names* are highly vulnerable to common obfuscation techniques. Also, excluded features such as *Manifest Receivers* generate very sparse features vectors, which effect the overall accuracy.

4.4.7   Feature Preprocessing

Feature extraction and similarity computation are the core operations in the proposed framework. Therefore, we need to optimize both their design and implementation to get the intended scalability. The expected output from *feature processing*

Table 4.1 Content feature categories

#	Content type features	Implemented feature
0	APK Byte N-grams	X
1	Classes.dex Byte N-grams	X
2	Native library bytes N-grams	X
3	Assembly Opcodes N-grams	X
4	Assembly class names	
5	Assembly method names	
6	Android API	X
7	Orphan android API	
8	Manifest permissions	X
9	Manifest activities	X
10	Manifest services	X
11	Manifest receivers	
12	IPs and URLs	X
13	APK files names	X
14	APK file light hashes (md4)	
15	Text strings	

is a vector, which can straightforwardly be used to compute the similarity between apps. App feature vectors are the input to Cypider community detection system.

The N-gram technique, presented in Sect. 4.4.1, suffers from its very high dimensionality D. The dimension number D hyper-parameter dramatically influences the computation and the memory needed by Cypider for Android malware detection. The complexity of computing the extracted N-grams features increases exponentially with N. For example, for the *opcodes N-grams*, described in Sect. 4.4.1, the dimension D equals to R^2 for bi-grams, where $R = 200$, the number of possible opcodes in Dalvik VM. Similarly, for *3-grams*, the dimension $D = R^3$; for *4-grams*, $D = R^4$. Furthermore, N has to be at least 3 or 5 to capture the semantics of some Android APK content.

To address this issue, we leverage the *hashing trick* technique [1] to reduce the high dimensionality of an arbitrary vector to a fixed-size feature vector. More formally, the *hashing trick* reduces a vector V with $D = R^N$ to a compressed version with $D = R^M$, where $M \ll N$. The compacted vector boosts Cypider, both computation-wise and memory-wise, by allowing the clustering system to handle a large volume of Android apps. Previous research [1, 8] has shown that the hash technique could preserve a decent amount of information in the vector distance. Moreover, the computational cost incurred by using the hashing technique for reducing dimensionality grows linearly with the number of samples and groups. Algorithm 1 illustrates the overall process of computing the compacted feature vector from an N-grams set. Furthermore, it helps to control the length of the compressed vector in an associated feature space.

4.5 LSH Similarity Computation

Building the *similarity network* is the backbone of Cypider framework. We generated the *similarity network* by computing the pairwise similarity between each feature vector of the apps APKs. As a result, we obtain multiple similarities according to the number of these content vectors. Using various similarities gives flexibility and modularity to Cypider. In other words, we could add any new feature vector to the similarity network without disturbing Cypider process. Also, we could remove features without affecting the overall process, which makes the experimentation of selecting the best features more convenient. More importantly, having multiple similarities between apps static contents in the similarity network leads to explainable decisions, where the investigator can track which contents a pair of apps are similar in the final similarity network. Similarity computation needs to be conducted in an efficient way that is much faster than the brute-force computation. For this purpose, we leverage LSH techniques, and more precisely *LSH Forest* [2], a tunable high-performance algorithm for similarity computation, employed by the Cypider framework. The key idea behind *LSH Forest* is that similar items hashed using *LSH* are most likely to be in the same bucket (collide) and dissimilar items will be in different ones. Many similarity measures correspond to *LSH* function with this property. In our case, we use the well-known *Euclidean* distance for this purpose.

$$d(m, n) = \|V_m - V_n\| = \sqrt{\sum_{i=1}^{|S|} (V_m(i) - V_n(i))^2} \qquad (4.1)$$

Given a pair of Android apps, after extracting one content feature vector, we use the *Euclidean* distance to compute the distance between two feature vectors m and n of one APK content, as depicted in Eq. 4.1. Figure 4.2 shows the LSH computation time with respect to the number of apps using *one CPU core* and *one thread* for the permission feature vector. Even though the current performance using *LSH Forest* is acceptable for a large number of daily malware samples (reaching 40,000 apps per hour), we believe that we could drastically improve these results by just leveraging an implementation that exploits all CPU cores in addition to multi-threading. The final result of the similarity computation is a *heterogeneous network*, where the nodes are the apps, and edges represent similarities between apps if a certain threshold is exceeded. The **similarity threshold** is the percentage of **average similarity** of a given content. In other words, we compute the average value of all the pairwise similarities for each feature content. Afterward, we set a percentage from this average to be the final threshold. We use **similarity threshold** for all feature contents even though they have different average values. The **similarity threshold** is systematically fixed based on our evaluation, and the same threshold is used in all experiments. However, we investigate the effect of the **similarity threshold** on Cypider performance in the evaluation section. Note that the network

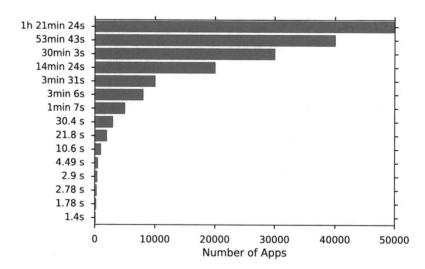

Fig. 4.2 LSH similarity computational time

is heterogeneous because there are multiple types of edges, where the edge type represents static content type.

4.6 Community Detection

A scalable community detection algorithm is essential to extract *suspicious communities*. For this reason, we empower **Cypider** with *Fast unfolding community detection* algorithm [3], which can scale to billions of network links. The algorithm achieves excellent results by measuring the *modularity* of communities. The *modularity* is a scalar value $M \in [-1, 1]$ that measures the density of edges inside a given community compared to the edges between communities. The algorithm uses an approximation of *modularity* since finding the exact value is computationally hard [3]. The previous algorithm requires a homogeneous network as input to work properly.

For this reason, we propose using a *majority-voting* mechanism (Sect. 3.1.2.3) to homogenize the heterogeneous network generated by similarity computation. Given the number of content similarity links s, the *majority-voting* method decides whether a pair of apps are similar or not by computing the ratio s/S, where S is the number of all contents used in the current **Cypider** configuration. If the ratio is above the average, the apps will only have one link in the *similarity network*. Otherwise, all the links will be removed. Notice that content similarity links could be retained for later use, for example, to conduct a thorough investigation about given apps to figure out how similar they are, and on which content they are similar. The prior use case could be of a great importance for security analysts.

We propose a *majority-voting* mechanism to filter links between the nodes (apps) to prevent having inaccurate *suspicious communities*. Furthermore, we employ a *degree filtering* hyper-parameter to filter all node links with a degree that is less than a threshold value. The previous hyper-parameter keeps only edges of a given node when their number is above the threshold. We call this hyper-parameter a **content threshold**, which is the number of similar contents to keep a link in the final similarity network.

Consequently, only nodes with high connectivity will maintain their edges, which are supposedly similar to malicious apps. Notice that all parameters have been fixed in our evaluations. In the case of a *mixed dataset* scenario, we use the degree 1 to filter all apps having a similarity link to themselves since they are not similar to any other app in the *active dataset*. **Cypider** filters these apps and consider them as benign apps. At this point, **Cypider** applies the community detection algorithm [3] to extract a set of communities with different sizes. Afterward, all communities are filtered with a community cardinal that is less than the *minimum community size* parameter (fixed for all the evaluations). The purpose of the filtrating is to prevent the extraction of bad quality communities. Figure 4.3 depicts an example of using **Cypider** on a small Android dataset (250 malware apps), where the process of community detection starts with *homogeneous network* and ends up with *suspicious*

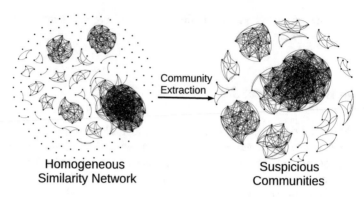

Fig. 4.3 Applying cypider on a small dataset

communities. The **content threshold** and **community size** hyper-parameters are thoroughly investigated in the evaluation section.

4.7   Community Fingerprint

Finding malicious communities is not the only goal of Cypider framework. Since Cypider is completely unsupervised, we aim at generating fingerprints from the extracted communities automatically. Therefore, in the following iteration, Cypider filters known apps without adding them to the *active dataset*. Traditional *cryptography fingerprints* or *fuzzy hashing* techniques are not suitable for our case since we aim to generate a fuzzy fingerprint, not only for one app but also for the whole malicious community whether it is a malware family or subfamily. In this context, we present a novel fingerprinting technique using the *One-Class Support Vector Machine* learning model (OC-SVM) [4], which was found to be fuzzy enough to cover malicious community apps. The *One-Class SVM* model is used to detect the set of a set of malicious apps of a given community. In particular, it detects the *soft boundary model* of that set. Adopting the *one-class model*, Cypider classifies new apps as belonging to this community or not. The proposed fingerprinting technique produces a much more compressed signature database compared to the traditional methods, where the signature is generated only for one malicious app. Moreover, it notably reduces the computation since we check only the community fingerprints instead of checking each single malware hash signature. In order to generate a *community fingerprint* from a set of malicious apps, Cypider first extracts the features, as presented in Sect. 4.4. Afterward, the *one-class model* is trained using static features of malicious community apps.

4.8 Experimental Results

In this section, we present **Cypider** implementation and the testing setup, including the dataset and the performance measurement techniques. Afterward, we present the achieved results regarding the defined metrics for both usage scenarios that are adopted in **Cypider** framework, namely *malware only* (only malicious apps) and *mixed* (benign and malicious apps) datasets.

4.8.1 Dataset and Test Setup

In order to evaluate **Cypider**, we leverage well-known Android datasets, namely (1) MalGenome malware dataset [9, 10], (2) Drebin malware dataset [11–13], and AndroZoo public Android app repository. As presented in Table 4.2, we produce two other evaluation datasets based on the previous ones by adding Android apps downloaded from Google Play in late 2014 and the beginning of 2019. These apps have been randomly downloaded without considering their popularity or any other factor. In order to build *Drebin Mixed* and *AndroZoo Mixed* datasets, we added 4403 benign apps to the original *Drebin* dataset. The result is a mixed dataset (malware and benign) with 50% of apps in each category. Similarly, we build the *MalGenome Mixed* dataset with 75% of benign apps.

The aim of using these datasets is to evaluate **Cypider** in unsupervised usage scenarios, with and without benign apps. First, we assess **Cypider** on malware only using *Drebin*, *AndroZoo*, and *Genome* datasets. This use case is the most attractive one in bulk malware analysis since it decreases the number of malware to be analyzed by considering only a sample from each detected community. Second, *Cypider* is evaluated against mixed datasets. The second scenario is more challenging because we expect not only *suspicious communities* as output but also benign communities (false positives) along with filtered benign apps.

To asses **Cypider** obfuscation resiliency, we conduct the evaluation on PRA-Gaurd obfuscation dataset,[1] which contains 11k obfuscated malicious apps using common obfuscation techniques [14]. In addition, we generate 100k benign and

Table 4.2 Evaluation datasets

	Drebin	DrebinMixed	Genome	GenomeMixed	AndroZoo	AndroZooMixed
Size	4330	8733	1168	4239	66k	66k
Malware	4330	4330	1168	1168	66k	66k
Benign	0	4403	0	3071	0	44k
Families	46	46	14	14	–	110k

---

[1]http://pralab.diee.unica.it/en/AndroidPRAGuardDataset.

malware obfuscated apps using DroidChameleon obfuscator [15] using common obfuscations techniques and related combinations.

To this end, various metrics are needed to measure **Cypider** performance in each dataset. We adopted the flowing metrics:

4.8.1.1   App Detection Metrics

A1: *True Malware*: This metric computes the number of malware apps that are detected by **Cypider**. It is applied to both usage scenarios.

A2: *False Malware*: This metric computes the number of benign apps that have been detected as malware apps. It is applied only to the *mixed dataset* since there are no benign apps in the other datasets.

A3: *True Benign*: This metric computes the number of filtered benign apps by **Cypider**. It is only applied to *mixed dataset* evaluation.

A4: *False Benign*: This metric computes the number of malware apps that are considered as benign in the *mixed dataset* evaluation.

A5: *Detection Coverage*: It measures the percentage of the detected malware from the overall dataset. Formally, it is the number of clustered Android apps divided by the total number of apps in the input dataset.

4.8.1.2   Community Detection Metrics

C1: *Detected Communities*: It indicates the number of suspicious communities that have been extracted by **Cypider**.

C2: *Pure Detected Communities*: This metric computes the number of communities with a unique Android malware family. In other words, a community is pure if it contains instances of the same family. In this task, we rely on the labels of the used datasets to check the purity of a given community. This metric is applied to both usage scenarios.

C3: *K-Mixed Communities*: This metric counts the communities with K-mixed malware families, where K is the number of families in a detected community. This metric is applied to both usage scenarios.

C4: *Benign Communities*: This metric computes the number of benign communities that have been detected as suspicious. This metric is applied to in the *mixed dataset* evaluation.

4.8.2   Mixed Dataset Results

Table 4.3 presents the evaluation results of **Cypider** using *Drebin Mixed* and *Genome Mixed* datasets. The most noticeable result is the fact that **Cypider** detects about *half* of the actual malware in a single iteration in both datasets even though

Table 4.3 Mixed evaluation
using apps metrics

Community metrics	Drebin mixed	Genome mixed
True Malware A1	2413	449
False Malware A2	190	103
True Benign A3	257	171
False Benign A4	38	10

Table 4.4 Evaluation using
community metrics

Apps metrics	Drebin mixed	Genome mixed
Detected C1	188	61
Pure detected C2	179	61
2-mixed C3	9	0
Benign C4	18	16

the noise of benign apps (false positive) is about 50–75% of the actual dataset. On
the other hand, **Cypider** is able to filter a considerable number of benign apps from
the dataset. However, in both dataset evaluations, we obtain some *false malware*
(190–103 apps) and *false benign* (38–10 apps), respectively, to datasets. According
to our results, these *false positives,* and *false negatives* appear, in most cases, in
communities with the same labels (malware or benign). Therefore, the investigation
would be straightforward by analyzing some samples from a given suspicious
community. The similarity network and the resultant communities are illustrated
in Fig. 4.5a and b, respectively.

Table 4.4 presents the results of **Cypider**'s evaluation using *community metrics.*
A very interesting result here is the number of *pure detected communities*, which
is 179 pure communities out of 188 detected communities in *Mixed Drebin* and
61 pure communities out of 61 detected communities (perfect purity) in *Mixed
Genome.* Consequently, almost all the detected communities have instances in the
same malware family or benign ones. Even the *mixed communities* are composed
of only two labels (2-mixed). It is important to mention that all the *detected benign
communities* are pure without any malware instance, which makes the investigation
much easier. Furthermore, according to our analysis, most malware labels in the
2-mixed malicious communities are just a naming variation of the same malware,
which is caused by name convention differences among vendors. For example, in
one *2-mixed* community, we found *FakeInstaller* and *Opfake* malware instances.
Actually, these names point to the same malware [16], which is *FakeInstaller.*
Similarly, we found *FakeInstaller* and *TrojanSMS.Boxer.AQ*, which points to the
same malware [17] with different vendor naming.

4.8.3 Results of Malware-Only Datasets

Tables 4.5 and 4.6 present the performance results of **Cypider** using the *app
metrics* and *community metrics* on malware only datasets. Since we use the same

Table 4.5 Malware
evaluation using apps metrics

Community metrics	Drebin	Genome
True Malware A1	2223	449

Table 4.6 Evaluation using
community metrics

Apps metrics	Drebin	Genome
Detected C1	170	45
Pure detected C2	161	45
2-mixed C3	9	0

malware dataset as the *mixed dataset* by only excluding benign apps, we obtain almost the same results. Cypider was able to detect about 50% of all malware in one iteration. Moreover, nearly all the recognized communities are pure. This high quality result is a significant advantage of Cypider in malware investigation since the security analyst could automatically attribute the family to a given suspicious community could be by only matching one or two samples. Furthermore, the analysis complexity dramatically decreases from 2413 detected malware to only 188 discovered communities. We believe that this could reduce the analysis window and help overcome the overwhelming number of daily detected Android malware. Notice that there are nine *2-mixed* communities in the *Drebin dataset*, which contain different malware labeled for the same actual malware, as mentioned before. Figure 4.4a depicts the *similarity network* of the Drebin malware dataset. After applying the community detection algorithm, we end up with *malicious communities*, as depicted in Fig. 4.4b.

4.8.4   Community Fingerprint Results

Table 4.7 shows the evaluation results with respect to *community fingerprinting*, which is applied to different detected communities with various Android malware families. The community fingerprint model (One-Class SVM) achieves 87% F1-score in detecting malware from the same malware family that is used in the training phase. In the *signature database*, these new malware samples share the same family with a given *community fingerprint*. Furthermore, the compressed format of this fingerprint, i.e., learning model in a binary format, could fingerprint an entire Android family, which generated a significantly more compacted *signature database*.

The performance of the community fingerprint mainly depends on the number of malware in the detected community. Higher detection performance is achieved when more malware instances exist in the community. In this respect, we determine a threshold systematically for the community cardinality (size), which is required to compute the fingerprint and store it in the *signature database*. As shown in Table 4.7, one of the main characteristics of community fingerprinting is its ability to differentiate between general malware and benign samples with high accuracy. The

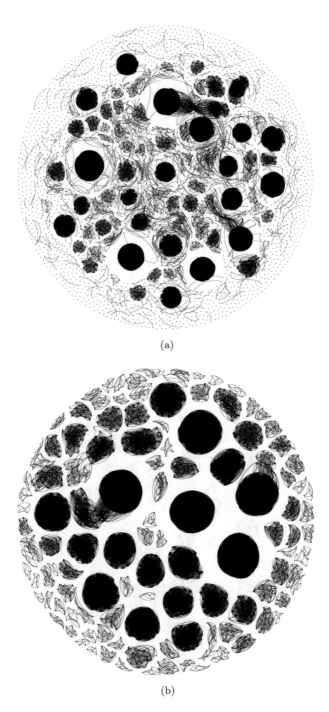

(a)

(b)

Fig. 4.4 Cypider network of Drebin malware dataset. (**a**) Similarity network. (**b**) Detected communities

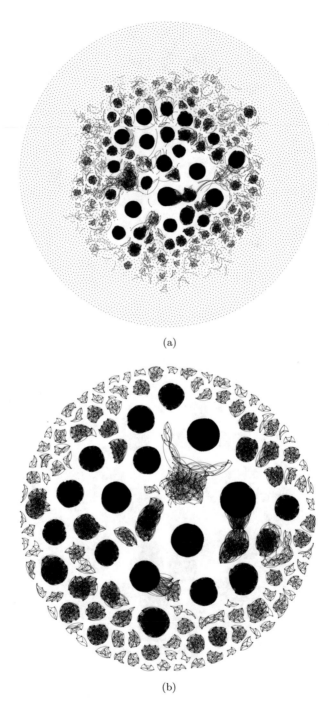

(a)

(b)

Fig. 4.5 Cypider network of Drebin malware dataset. (**a**) Simialrity network. (**b**) Detected communities

Table 4.7 Community fingerprint accuracy on different families

Family	Community size	Family detection (Acc)	Malware (Acc)	Benign (Acc)	General (F1)
Hidden Ads	1774	77.43	93.40	94.93	87.83
Bridge Master	1041	76.99	93.08	93.22	87.05
Info Stealer	2973	86.60	88.37	91.26	75.50
Plankton	495	77.76	100.0	100.0	75.10
Base Bridge	1499	76.76	88.51	92.47	73.28
Utchi	973	76.19	99.98	100.0	72.02

reason behind this high accuracy is the high similarity between general malware and the trained family. Notice that the one-class SVM model is trained on samples from only one malware family. In other words, malicious apps tend to have similar features, although they do not belong to the same malware family. Thus, benign samples and general malware are highly dissimilar, and hence benign samples are less likely to match with community fingerprint, which minimizes the overall false positive.

4.9 Hyper-Parameter Analyses

In this section, we analyze the effect of the employed hyper-parameters on Cypider on the overall performance measured using **Purity, Coverage**, and **Community Numbers** metrics. Specifically, we investigate the **similarity threshold**, the **content threshold**, and the **community size**, as presented in Sects. 4.5 and 4.6.

4.9.1 Purity Analysis

In the purity analysis, we compute the overall percentage of clustered malware samples of the groups belonging to the same Android malware family. A perfect purity metric means that each detected community (cluster) contains samples from the same Android malware family. Figures 4.6 and 4.7 show the effect of Cypider hyper-parameters on the purity of the detected malware communities in the similarity network of *Drebin* and AndroZoo datasets, respectively.

It is worth noting that the **content threshold** is the most affecting hyper-parameter on the overall purity. A small content threshold results in a lower purity percentage, as shown in the evaluation of both *Drebin* and *AndroZoo* datasets. This

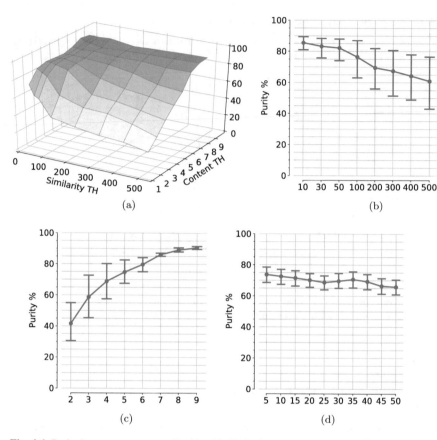

Fig. 4.6 Purity hyper-parameters on Drebin. (**a**) Similarity and content. (**b**) Similarity threshold. (**c**) Content threshold. (**d**) Community size

finding is intuitive because **Cypider** grouping outcome is more accurate when using more content types threshold in the *majority-voting* similarity computation.

On the other hand, the **similarity threshold** has a secondary effect compared to the **content threshold**. This means that a tight threshold outputs less false samples in the detected communities. Finally, we notice a very minor effect of the **community size** on the overall purity metric for both *Drebin* and *AndroZoo* evaluations, as shown in Figs. 4.6 and 4.7, respectively.

4.9.2   Coverage Analysis

In the coverage analysis, we assess the percentage of the detected malware from the overall input dataset. A perfect coverage means that **Cypider** detects malware in the produced malware communities. Figures 4.8 and 4.9 depict the change in

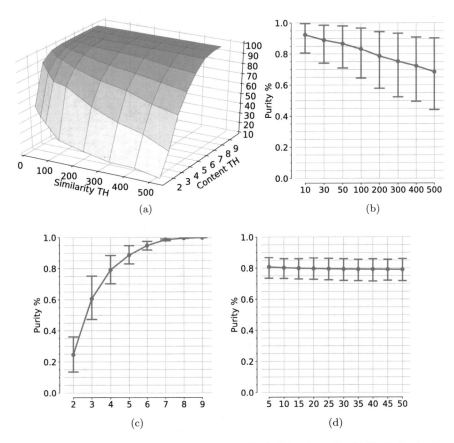

Fig. 4.7 Purity hyper-parameters on AndroZoo. (**a**) Similarity and content. (**b**) Similarity threshold. (**c**) Content threshold. (**d**) Community size

the coverage percentage with **Cypider** hyper-parameters for *Drebin* and *AndroZoo* datasets, respectively.

We notice that the **content threshold** is the most affecting hyper-parameter on the overall coverage metric. This means that a high content threshold in the majority voting (Sect. 4.5) leads to the detection of fewer malware samples in the produced malware communities. Therefore, the coverage metric decreases drastically with a high content threshold, as shown in Figs. 4.8 and 4.9.

The **similarity threshold** and the **community size** have a secondary effect on the coverage metric. For the similarity threshold, a wide distance threshold yields a higher detection rate, and therefore a high coverage metric. For the community size threshold, a larger value leads to ignore many small malware communities, which affects the detection coverage metric negatively.

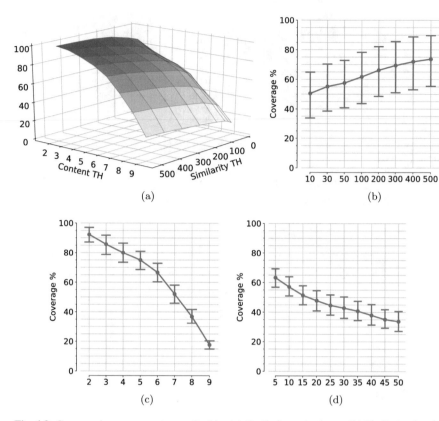

Fig. 4.8 Coverage hyper-parameters on Drebin. (**a**) Similarity and content. (**b**) Similarity threshold. (**c**) Content threshold. (**d**) Community size

4.9.3 Number of Communities Analysis

In this section, we analyze the total number of the detected communities produced by **Cypider**. A perfect **Cypider** clustering yields a result in which the number of communities is equal to the actual number of malware families in the input dataset. Figures 4.10 and 4.11 depict the effect of **Cypider** hyper-parameters on the number of the detected communities on *Drebin* and *AndroZoo* datasets, respectively.

It is essential to mention that the community size has a strong influence on the number of communities. A higher community size threshold will filter many small malware communities, which influences the number of detected communities directly.

For the content threshold, the majority voting with a small content threshold causes many communities to merge. This is because the samples have to be similar in only two content thresholds to maintain a similarity. On the other hand, the majority voting with a high content threshold will detect fewer malware samples

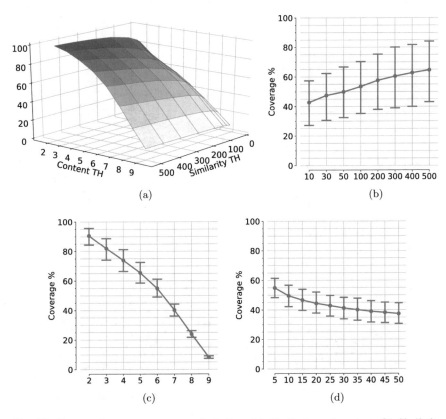

(a)   (b)

(c)   (d)

Fig. 4.9 Coverage hyper-parameters on AndroZoo. (**a**) Similarity and content. (**b**) Similarity threshold. (**c**) Content threshold. (**d**) Community size

and, hence, less overall malicious communities. Finally, we notice a minor effect of the similarity threshold on the overall number of the detected communities, as depicted in Figs. 4.10 and 4.11 for *Drebin* and *AndroZoo* datasets, respectively.

4.9.4 Efficiency Analysis

In this section, we investigate the overall efficiency of the Cypider framework. Specifically, we present the runtime in seconds on the core computation of our framework: (1) the **similarity computation** to build the similarity network, (2) the **community detection** to partition the similarity network into a set of malicious communities.

Figure 4.12 depicts the similarity computation time in seconds to build Cypider similarity network. We notice that Cypider framework is very efficient at producing the similarity network because we employ locality sensitive hashing techniques to

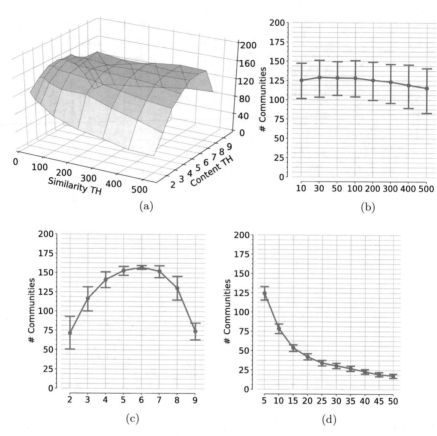

Fig. 4.10 Detected communities hyper-parameters on Drebin. (**a**) Similarity and content. (**b**) Similarity threshold. (**c**) Content threshold. (**d**) Community size

speed up the pairwise similarity computation between the feature hashing vectors. For example, **Cypider** took only about 1 s to compute the similarity between $111k$ samples feature hashing vectors.

Figures 4.13 and 4.14 present the community detection time in seconds on *Drebin* and *AndroZoo* datasets, respectively. We analyze the effect of *similarity* and *content* thresholds on the overall community detection time. In Figs. 4.13 and 4.14, we notice that **Cypider** spends more time due to decreasing the content threshold and decreasing the similarity threshold in *Drebin* and *AndroZoo* experiments. The previous thresholds setup increases the density of **Cypider** similarity network, and therefore, the community detection processing takes more time in the partition process of the network. On the other hand, increasing the content threshold while decreasing the similarity threshold produces a very sparse similarity network, which takes a negligible time in the partition process.

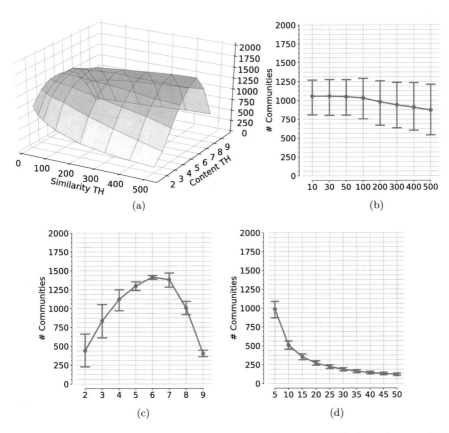

Fig. 4.11 Detected communities hyper-parameters on AndroZoo. (**a**) Similarity and content. (**b**) Similarity threshold. (**c**) Content threshold. (**d**) Community size

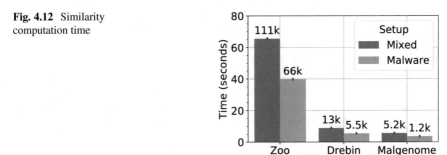

Fig. 4.12 Similarity computation time

4.10   Case Study: Recall and Precision Settings

In this section, we evaluate **Cypider** framework with respect to recall and precision. We aim to assess **Cypider** performance in terms of *purity* and *coverage* in case the security practitioner focuses on having: (1) maximum recall (a minimum

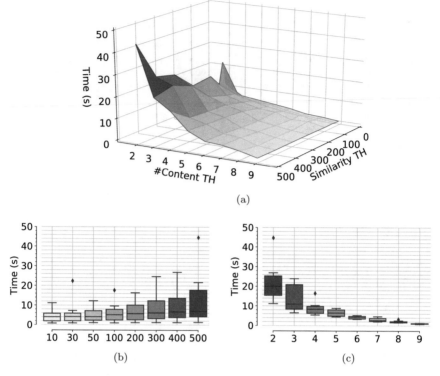

Fig. 4.13 Community detection time analysis on Drebin. (**a**) Similarity and content. (**b**) Similarity threshold. (**c**) Content threshold

false detection), or (2) maximum precision (maximum coverage). Both recall and precision settings are common in real deployments. We tune the recall and precision settings by adjusting **Cypider** hyper-parameters to reach the set goal.

Figure 4.15 presents **Cypider** malware only performance on different datasets (*MalGenome, Drebin, AndroZoo*) under recall and precision settings. In the recall setting, **Cypider** achieves 95–100% purity while maintaining 15–26% malware coverage. Therefore, we detect about 20% (on average) of the input malware in form of communities with 98% purity (Fig. 4.15, recall charts). On the other hand, **Cypider** achieves 63–73% malware coverage while maintaining 84–93% purity, as shown in Fig. 4.15 (precision charts).

The contrast between the recall and precision settings is more clearly visible in the similarity network, as shown in Figs. 4.16 and 4.17 for *MalGenome* and *Drebin* datasets, respectively. The aforementioned figures present each malware family in a different color. Malware communities depicted with more than one color contain more than one malware family. Pure malware communities have only one color in the edges and nodes. We notice more detected malware communities in the similarity network in the precision settings. In contrast, in the recall similarity

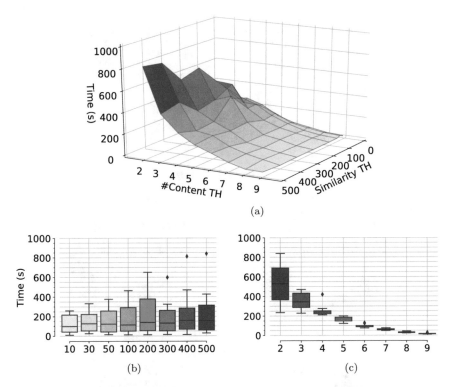

Fig. 4.14 Community detection time analysis on AndroZoo. (**a**) Similarity and content. (**b**) Similarity threshold. (**c**) Content threshold

network, we notice fewer malware communities, and most of the nodes are part of any community (not detected).

Figure 4.18 depicts **Cypider** mixed performance under recall and precision settings for *MalGenome, Drebin*, and *AndroZoo* datasets. The most noticeable result is that all the detected benign communities have perfect purity metrics under both recall and precision settings. Moreover, benign coverage is less than the malware coverage under all settings. In other words, **Cypider** could bring benign samples during clustering but gathered in pure communities, which is very helpful in case of manual investigations.

The difference between recall and precision settings in the mixed scenario is more clearly visible in the similarity networks. Figures 4.19 and 4.20 show the recall and precision similarity network of *MalGenome* and *Drebin* datasets, respectively. Darker color communities contain malware samples, and lighter color communities contain benign samples. We notice a clear separation between malicious communities and benign ones. Also, more numerous and larger communities have been detected under the precision setting compared to the recall setting.

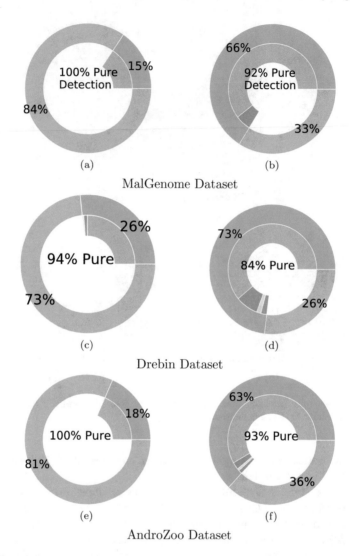

Fig. 4.15 Performance under recall/precision settings. (**a**) Recall. (**b**) Precision. (**c**) Recall. (**d**) Precision. (**e**) Recall. (**f**) Precision

Tables 4.8 and 4.9 detail **Cypider** performance under the recall and the precision settings in terms of coverage/purity and number of detected/pure communities, respectively.

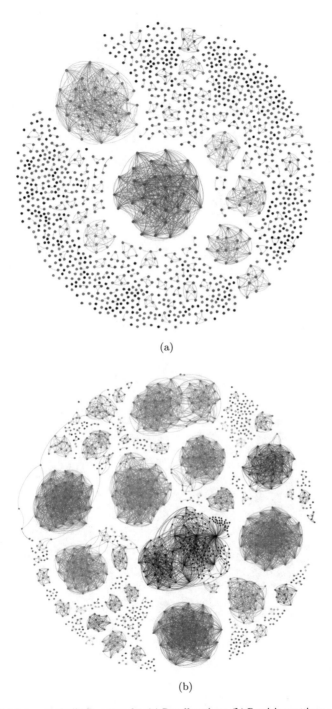

(a)

(b)

Fig. 4.16 Malgenome similarity networks. (**a**) Recall settings. (**b**) Precision settings

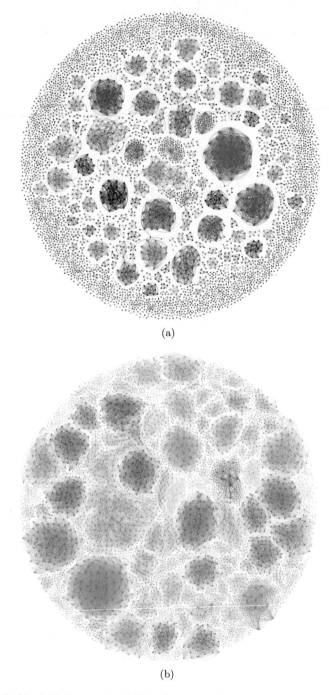

(a)

(b)

Fig. 4.17 Drebin similarity networks. (**a**) Recall settings. (**b**) Precision settings

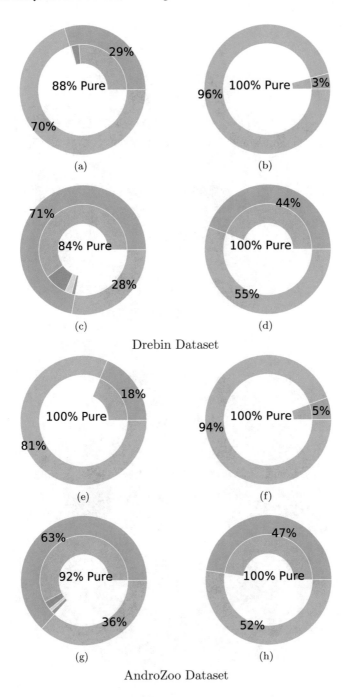

Fig. 4.18 Performance under mixed recall/precision settings. (**a**) Malware (Recall). (**b**) Benign (Recall). (**c**) Malware (Precision). (**d**) Benign (Precision). (**e**) Malware (Recall). (**f**) Benign (Recall). (**g**) Malware (Precision). (**h**) Benign (Precision)

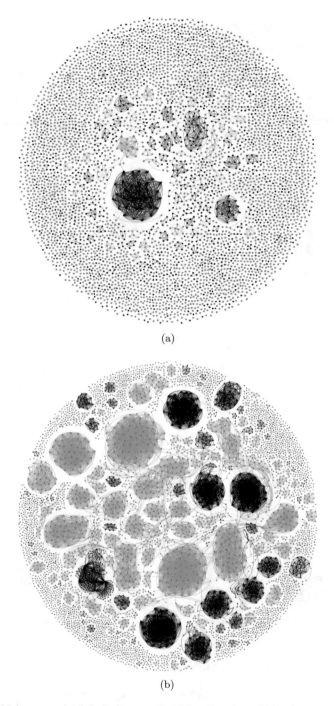

(a)

(b)

Fig. 4.19 Malgenome mixed similarity network. (**a**) Recall settings. (**b**) Precision settings

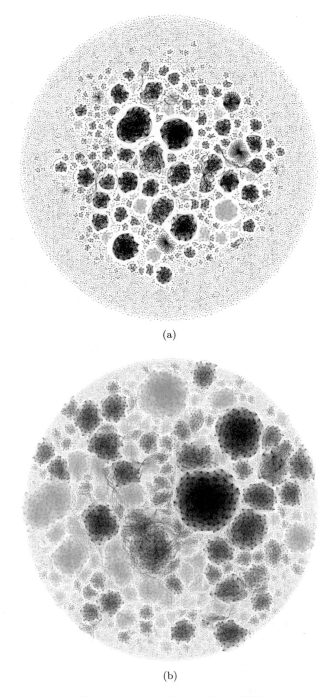

(a)

(b)

Fig. 4.20 Drebin mixed similarity network. (**a**) Recall settings. (**b**) Precision settings

Table 4.8 Coverage and purity details

Evaluation setup			Dataset size			Coverage/purity %		
Dataset	Scenario	Settings	#Benign	#Malware	#Total			
Malgenome	Malware	*Recall*	/	1.23k	1.23k	/	15.56%/99.0%	15.56%/99.0%
		Precision	/	1.23k	1.23k	/	66.21%/91.0%	66.21%/91.0%
	Mixed	*Recall*	4.03k	1.23k	5.26k	3.7%/100.0%	15.48%/99.0%	6.46%/100.0%
		Precision	4.03k	1.23k	5.26k	39.66%/100.0%	61.02%/92.0%	44.68%/98.0%
Drebin	Malware	*Recall*	/	5.55k	5.55k	/	26.69%/93.0%	26.69%/93.0%
		Precision	/	5.55k	5.55k	/	73.08%/84.0%	73.08%/84.0%
	Mixed	*Recall*	7.64k	5.55k	13.19k	3.85%/100.0%	29.77%/88.0%	14.74%/90.0%
		Precision	7.64k	5.55k	13.19k	44.27%/100.0%	72.0%/83.0%	55.93%/91.0%
AndroZoo	Malware	*Recall*	/	66.76k	66.76k	0.0%/0.0%	18.48%/100.0%	18.48%/100.0%
		Precision	/	66.76k	66.76k	0.0%/0.0%	63.03%/92.0%	63.03%/92.0%
	Mixed	*Recall*	44.18k	66.76k	110.94k	5.52%/100.0%	18.7%/100.0%	13.45%/100.0%
		Precision	44.18k	66.76k	110.94k	47.63%/100.0%	63.19%/91.0%	56.99%/94.0%

Table 4.9 Number of detected/pure communities details

Evaluation setup			Dataset size			#Communities/#Pure		
Dataset	Scenario	Settings	#Benign	#Malware	#Total	Benign	Malware	Overall
Malgenome	Malware	*Recall*	/	1.23k	1.23k	/	15/15	15/15
		Precision	/	1.23k	1.23k	/	35/31	35/31
	Mixed	*Recall*	4.03k	1.23k	5.26k	18/18	17/17	35/35
		Precision	4.03k	1.23k	5.26k	71/71	34/31	105/102
Drebin	Malware	*Recall*	/	5.55k	5.55k	/	95/89	95/89
		Precision	/	5.55k	5.55k	/	155/136	155/136
	Mixed	*Recall*	7.64k	5.55k	13.19k	30/30	109/102	139/132
		Precision	7.64k	5.55k	13.19k	125/125	152/132	277/257
AndroZoo	Malware	*Recall*	/	66.76k	66.76k	/	800/798	800/798
		Precision	/	66.76k	66.76k	/	1355/1291	1355/1291
	Mixed	*Recall*	44.18k	66.76k	110.94k	176/176	828/826	1004/1002
		Precision	44.18k	66.76k	110.94k	586/586	1321/1250	1907/1836

4.11 Case Study: Obfuscation

In this section, we investigate the robustness of **Cypider** framework against common obfuscation techniques and code transformation in general. We employ *PRAGuard* obfuscated Android malware, which contains $11k$ samples, along with benign samples from *AndroZoo* dataset. Table 4.10 details **Cypider** performance on the malware and the mixed scenarios. We compare **Cypider** performance before and after applying a single or a combination of obfuscation techniques, as shown in Table 4.10.

The evaluation results show that common obfuscation techniques have a limited effect on **Cypider** performance in general (60–77% coverage and 71–99% purity). *Class encryption* obfuscation decreases the coverage from 66% in the non-obfuscated dataset to 36–38%. However, *Class encryption* does not affect the purity. Similarly, *Reflection* obfuscation technique dropped down the purity to 67–71% compared to the original dataset but does not affect the coverage performance. To strengthen our findings (Table 4.10) on *PRAGuard* obfuscation dataset, we build our obfuscation dataset using *DroidChameleon* obfuscation tool. We obfuscate *Drebin* malware dataset ($5k$ malware samples) and benign samples from *AndroZoo* dataset ($5k$ malware samples). The result is $100k$ samples ($50k$ malware and $50k$ benign) from different obfuscation settings, as shown in Table 4.11. Similar to *PRAGuard* experiment, we compare **Cypider** performance before obfuscation (original *Drebin* dataset) and after obfuscation. However, this experiment is different from the *PRAGuard* one because both benign and malicious samples are obfuscated.

Table 4.11 details the result of **Cypider** framework on the different obfuscation techniques. The most noticeable is that the obfuscation techniques provided by the *DroidChameleon* tool have a limited effect on the clustering. All the performance metrics remain stable on both non-obfuscated and obfuscated samples under the

Table 4.10 Performance on obfuscated—PRAGaurd dataset

Evaluation setup		Coverage/purity %			#Communities/#Pure		
Scenario	Obfuscation	Bengin	Malware	Overall	Bengin	Malware	Overall
Malware	Malgenome (original)	/	66.2%/92.3%	66.2%/92.3%	/	35/31	17/17
	(1) TRIVIAL	/	60.3%/99.8%	60.3%/99.8%	/	34/34	35/31
	(2) STRING ENCRYPTION	/	63.5%/96.7%	63.5%/96.7%	/	34/32	35/31
	(3) REFLECTION	/	70.8%/71.9%	70.8%/71.9%	/	30/27	35/31
	(4) CLASS ENCRYPTION	/	38.3%/98.5%	38.3%/98.5%	/	27/26	35/31
	(1) & (2)	/	52.4%/99.8%	52.4%/99.8%	/	42/42	35/31
	(1) & (2) & (3)	/	65.1%/67.5%	65.1%/67.5%	/	38/34	35/31
	(1) & (2) & (3) & (4)	/	36.3%/99.7%	36.3%/99.7%	/	39/39	35/31
Mixed	Malgenome (Orignal)	52.84%/100.0%	45.44%/95.0%	51.1%/99.0%	63/63	36/32	99/95
	(1) TRIVIAL	50.98%/100.0%	65.72%/90.0%	54.3%/97.0%	66/66	38/32	104/98
	(2) STRING ENCRYPTION	52.47%/100.0%	68.41%/93.0%	56.06%/98.0%	66/66	33/30	99/96
	(3) REFLECTION	51.45%/100.0%	77.23%/63.0%	57.21%/89.0%	65/65	29/21	94/86
	(4) CLASS ENCRYPTION	52.84%/100.0%	45.44%/95.0%	51.1%/99.0%	63/63	36/32	99/95
	(1) & (2)	/	/	/	/	/	/
	(1) & (2) & (3)	/	/	/	/	/	/
	(1) & (2) & (3) & (4)	49.04%/100.0%	44.64%/94.0%	48.01%/99.0%	66/66	39/37	105/103

Table 4.11 Performance on obfuscation—Drebin dataset

Evaluation setup		Coverage/purity %			#Communities/#pure		
Scenario	Obfuscation	Bengin	Malware	Overall	Bengin	Malware	Overall
Mixed	Drebin (Original)	44.27%/100.0%	72.0%/83.0%	55.93%/91.0%	125/125	152/132	277/257
	Class renaming	41.71%/100.0%	72.61%/83.0%	54.64%/91.0%	129/129	156/135	285/264
	Method renaming	42.02%/100.0%	71.08%/83.0%	54.19%/91.0%	121/121	149/129	270/250
	Field renaming	43.59%/100.0%	72.01%/83.0%	55.49%/91.0%	128/128	148/127	276/255
	Code reordering	43.43%/100.0%	71.49%/83.0%	55.19%/91.0%	127/127	155/135	282/262
	Debug information removing	44.34%/100.0%	72.09%/83.0%	55.96%/91.0%	117/117	151/130	268/247
	Junk code insertion	40.71%/100.0%	71.81%/83.0%	53.73%/90.0%	124/124	153/132	277/256
	Instruction insertion	42.65%/100.0%	71.25%/83.0%	54.63%/91.0%	120/120	156/136	276/256
	String encryption	43.2%/100.0%	72.16%/83.0%	55.32%/91.0%	133/133	147/127	280/260
	Array encryption	43.55%/100.0%	71.93%/83.0%	55.42%/91.0%	125/125	152/131	277/256
Malware	Drebin (Original)	/	73.08%/84.0%	73.08%/84.0%	/	155/136	155/136
	Class renaming	/	74.14%/84.0%	74.14%/84.0%	/	160/137	160/137
	Method renaming	/	72.66%/83.0%	72.66%/83.0%	/	159/136	159/136
	Field renaming	/	73.75%/83.0%	73.75%/83.0%	/	155/132	155/132
	Code reordering	/	74.07%/83.0%	74.07%/83.0%	/	158/135	158/135
	Debug information removing	/	72.92%/83.0%	72.92%/83.0%	/	155/132	155/132
	Junk code insertion	/	73.86%/83.0%	73.86%/83.0%	/	157/135	157/135
	Instruction insertion	/	73.96%/85.0%	73.96%/85.0%	/	160/137	160/137
	String encryption	/	73.8%/83.0%	73.8%/83.0%	/	155/132	155/132
	Array encryption	/	73.8%/83.0%	73.8%/83.0%	/	155/133	155/133

malware and mixed scenarios. We argue that Cypider framework is resilient to common obfuscation and code transformation techniques because our framework considers many APK contents for feature extraction. Therefore, the obfuscation techniques can affect one APK content, but Cypider is able to leverage other contents to fingerprint malware sample and compute the similarity with other malware samples.

4.12 Summary

In this chapter, we have leveraged APK-DNA fingerprint, proposed in the previous chapter, to design and implement an innovative, efficient and scalable framework for Android malware detection, called Cypider for Android malware clustering. In essence, the detection mechanism relies on the community concept. Cypider provides a systematic framework that can generate a fingerprint for each community, enabling the identification of known and unknown malicious communities. Cypider has been implemented and evaluated on different malicious and mixed datasets. Our findings show that Cypider is a valuable and promising framework for the detection of malicious communities. Cypider only needs few seconds to build a network similarity of a large number of apps. The community fingerprinting results are very promising as 87% of the detection is achieved.

In the next chapter, we present another fuzzy fingerprinting approach for Android malware detection with the following specifications: (1) We rely on dynamic analysis instead of static analysis. (2) We build on top the dynamic analysis fingerprints a classification system using supervised machine learning instead of the clustering approach employed in this chapter.

References

1. Q. Shi, J. Petterson, G. Dror, J. Langford, A.J. Smola, S.V.N. Vishwanathan, Hash kernels for structured data. J. Mach. Learn. Res. **10**, 2615–2637 (2009)
2. M. Bawa, T. Condie, P. Ganesan, LSH forest: Self-tuning indexes for similarity search, in *Proceedings of the 14th International Conference on World Wide Web, WWW 2005, Chiba, May 10–14, 2005* (2005), pp. 651–660
3. V.D. Blondel, J.-L. Guillaume, R. Lambiotte, E. Lefebvre, Fast unfolding of communities in large networks. J. Stat. Mech. Theory Exp. **2008**, P10008 (2008)
4. B. Schölkopf, J.C. Platt, J. Shawe-Taylor, A.J. Smola, R.C. Williamson, Estimating the support of a high-dimensional distribution. Neural Comput. **13**(7), 1443–1471 (2001)
5. The Android Native Development Kit (NDK), https://developer.android.com/ndk/index.html (2016). Accessed January 2016
6. L. Deshotels, V. Notani, A. Lakhotia, Droidlegacy: Automated familial classification of android malware, in *Proceedings of the 3rd ACM SIGPLAN Program Protection and Reverse Engineering Workshop 2014, PPREW 2014, January 25, 2014, San Diego* (2014), pp. 3:1–3:12

7. H.C. Wu, R.W.P. Luk, K. Wong, K. Kwok, Interpreting TF-IDF term weights as making relevance decisions. tACM Trans. Inf. Syst. **26**(3), 13:1–13:37 (2008)
8. K.Q. Weinberger, A. Dasgupta, J. Langford, A.J. Smola, J. Attenberg, Feature hashing for large scale multitask learning, in *Proceedings of the 26th Annual International Conference on Machine Learning, ICML 2009, Montreal, Quebec, June 14–18, 2009* (2009), pp. 1113–1120
9. Android Malware Genome Project, http://www.malgenomeproject.org/ (2015), Accessed January 2015
10. Y. Zhou, X. Jiang, Dissecting android malware: Characterization and evolution, in *IEEE Symposium on Security and Privacy, SP 2012, 21–23 May 2012, San Francisco* (2012), pp. 95–109
11. Drebin Android Malware Dataset, https://user.informatik.uni-goettingen.de/~darp/drebin/ 2015. Accessed January 2015
12. D. Arp, M. Spreitzenbarth, M. Hubner, H. Gascon, K. Rieck, DREBIN: effective and explainable detection of android malware in your pocket, in *21st Annual Network and Distributed System Security Symposium, NDSS 2014, San Diego, February 23–26, 2014* (2014)
13. M. Spreitzenbarth, F.C. Freiling, F. Echtler, T. Schreck, J. Hoffmann, Mobile-sandbox: having a deeper look into android applications, in *Proceedings of the 28th Annual ACM Symposium on Applied Computing, SAC '13, Coimbra, March 18–22, 2013* (2013), pp. 1808–1815
14. D. Maiorca, D. Ariu, I. Corona, M. Aresu, G. Giacinto, Stealth attacks: An extended insight into the obfuscation effects on android malware. Comput. Secur. **51**, 16–31 (2015)
15. V. Rastogi, Y. Chen, X. Jiang, Droidchameleon: evaluating android anti-malware against transformation attacks, in *8th ACM Symposium on Information, Computer and Communications Security, ASIA CCS '13, Hangzhou, May 08–10, 2013* (2013), pp. 329–334
16. Andr/OpFake-U malware family, https://www.sophos.com/en-us/threat-center/threat-analyses/viruses-and-spyware/Andr~OpFake-U.aspx (2016). Accessed February 2016
17. Andr/Boxer-C malware family, https://www.sophos.com/en-us/threat-center/threat-analyses/viruses-and-spyware/Andr~Boxer-C.aspx (2016). Accessed on May 2016

Chapter 5
Android Malware Fingerprinting Using Dynamic Analysis

In this chapter, we elaborate a data driven framework for detecting Android malware using automatically engineered features derived from dynamic analyses. The state-of-the-art solutions, such as [5, 12, 13], rely on manual feature engineering in malware detection. For example, StormDroid [5] uses *Sendsms* and *Recvnet* dynamic features, which are chosen based on statistical analysis, for Android malware detection. As another example, the authors in [14] used specific features to build behavioral graphs for Win32 malware detection. The security features may change based on the execution environment despite the target platform. For instance, the authors in [5] and in [15] used different security features due to the difference between the execution environments. In the context of a security application, we are looking for a portable framework for malware detection based on the behavioral reports across a variety of platforms, architectures, and execution environments. The security analyst would be able to rely on this plug-and-play framework with a minimum effort in terms of feature engineering. We plug the behavioral analysis reports for the training. Afterward, we employ the produced classification model on new reports without an explicit security feature engineering as in [5, 14, 16]. This previous process works virtually on any behavioral reports.

We propose, MalDy, a portable framework for malware detection and family threat investigation based on behavioral reports. MalDy framework is built on top of Natural Language Processing (NLP) modeling and supervised machine learning techniques. The main idea is to formalize a behavioral report, agnostic to the execution environment, into a Bag of Words (BoW) where the features are the reports' words. Afterward, we leverage machine learning techniques to automatically discover relevant security features that help differentiate and attribute malware. The result is MalDy, a portable (Sect. 5.5.3), effective (Sect. 5.5.2), and efficient (Sect. 5.5.4) framework for malware analysis.

5.1   Threat Model

We position MalDy as a generic malware analysis tool. MalDy considers only behavioral reports generated from the execution of program binary inside a sandboxing environment. Therefore, MalDy is by design resilient to binary code static analysis transformation like packing, compression, and dynamic loading. MalDy performance depends on the quality of the collected reports. The more security information and features are provided about malware samples in the reports, the more accurate MalDy could differentiate malware from benign and attribute to known families. The malware execution time and the random event generator of the sandboxing may have a considerable impact on MalDy because they affect the quality of the behavioral reports. Anti-emulation techniques, used to evade dynamic analysis, could be challenging for MalDy framework. However, this issue is related to the choice of the underlying execution environment.

5.2   Overview

The execution of a binary sample (or app) produces textual logs, whether in a controlled environment (software sandbox) or production ones. The execution logs are composed of a sequence of statements, as the result of the app execution events. Furthermore, each statement is a sequence of words that gives a more granular description of an actual app event. From a security analysis perspective, app behaviors are summarized in an execution report, which is a sequence of statements, and each statement is a sequence of words. Malicious apps tend to have distinguishable behaviors from benign apps, and this difference is translated into words in the behavioral report. Also, similar malicious apps (same malware family) behaviors tend to correspond to related words.

Nowadays, there are many software sandbox solutions for malware investigations. CWSandbox (2006–2011) was one of the first sandbox solutions for production use. It is presently known as ThreatAnalyzer,[1] owned by ThreatTrack Security. TheatAnalyzer is a sandbox system for Win32 malware, and it produces behavioral reports that cover most of malware behavioral aspects such as a file, network, and register access records. Figure 5.1 shows a snippet from a behavioral report generated by ThreatAnalyzer. For Android malware, we use *DroidBox* [1], a well-established sandbox environment based on Android software emulator [2] provided by Google Android SDK [3]. Figure 5.2 shows a snippet of a behavioral report generated using DroidBox.

---

[1]https://www.threattrack.com/malware-analysis.aspx.

```
<open_key~key="HKEY_LOCAL_MACHINE\Software\Microsoft\Windows
NT\CurrentVersion\AppCompatFlags\Layers"/> <open_key
key="HKEY_CURRENT_USER\Software\Microsoft\Windows
NT\CurrentVersion\AppCompatFlags\Layers"/> <open_key
key="HKEY_LOCAL_MACHINE\System\CurrentControlSet\Services\
LanmanWorkstation\NetworkProvider"/>
</registry_section> <process_section> <enum_processes
apifunction="Process32First" quantity="84"/> <open_process targetpid="308"
desiredaccess="PROCESS_ALL_ACCESS PROCESS_CREATE_PROCESS PROCESS_CREATE_THREAD
PROCESS_DUP_HANDLE PROCESS_QUERY_INFORMATION PROCESS_SET_INFORMATION
PROCESS_TERMINATE PROCESS_VM_OPERATION PROCESS_VM_READ PROCESS_VM_WRITE
PROCESS_SET_SESSIONID PROCESS_SET_QUOTA SYNCHRONIZE"
apifunction="NtOpenProcess" successful="1"/>
```

Fig. 5.1 Win32 behavioral report

```
"accessedfiles": { "1546331488": "/proc/1006/cmdline","2044518634":
"/data/com.macte.JigsawPuzzle.Romantic/shared_prefs/com.apperhand.global.xml",
"296117026":
"/data/com.macte.JigsawPuzzle.Romantic/shared_prefs/com.apperhand.global.xml",
"592194838": "/data/data/com.km.installer/shared_prefs/TimeInfo.xml",
"956474991": "/proc/992/cmdline"},"apkName": "fe3a6f2d4c","closenet":
{},"cryptousage": {},"dataleaks": {},"dexclass": { "0.2725639343261719": {
    "path": "/data/app/com.km.installer-1.apk", "type": "dexload"}
```

Fig. 5.2 Android behavioral report

5.2.1 Notation

- $X = \{X_{build}, X_{test}\}$: X is the global dataset used to build and report MalDy performance in various tasks. We use the build dataset X_{build} to train and tune the hyper-parameters of MalDy models. The test set X_{test} is used to measure the final performance of MalDy, which is reported in the evaluation section. X is divided randomly and equally to X_{build} (50%) and X_{test} (50%). To build the sub-datasets, we employ the stratified random split on the main dataset.
- $X_{build} = \{X_{train}, X_{valid}\}$: Build set, X_{build}, is composed of the training set and validation set and used to build MalDy ensembles.
- $m_{build} = m_{train} + m_{valid}$: Build size is the total number of reports used to build MalDy. The training set takes 90% of the build dataset, and the rest is used as a validation set.
- $X_{train} = \{(x_0, y_0), (x_1, y_1), \ldots, (x_{m_{train}}, y_{m_{train}})\}$: The training set, X_{train}, is the training dataset of MalDy machine learning models.
- $m_{train} = |X_{train}|$: The size of m_{train} is the number of reports in the training set.
- $X_{valid} = \{((x_0, y_0), (x_1, y_1), \ldots, (x_{m_{valid}}, y_{m_{valid}})\}$: The validation set, X_{valid}, is the dataset used to tune the trained model. We choose the hyper-parameters that achieve the best scores on the validation set.

- $m_{valid} = |X_{valid}|$: The size of m_{valid} is the number of reports in the validation set.
- (x_i, y_i) : A single record in X is composed of a single report x_i and its label $y_i \in \{+1, -1\}$. The label meaning depends on the investigation task. In the detection task, a positive value means malware, and a negative means benign. In the family attribution task, a positive means the sample is part of the current model malware family, and a negative means is not.
- $X_{test} = \{((x_0, y_0), (x_1, y_1), .., (x_{m_{test}}, y_{m_{test}}))\}$: We use X_{test} to compute and report back the final performance results as presented in the evaluation section (Sect. 5.5).
- $m_{test} = |X_{test}|$: m_{test} is the size of the X_{test} and it represents 50% of the global dataset X.

5.3   Methodology

In this section, we present the general approach of MalDy, as illustrated in Fig. 5.3. The section describes the approach based on the chronological order of the building steps.

5.3.1   Behavioral Reports Generation

MalDy Framework starts from a dataset X of behavioral reports with known labels (malware or benign labels for the detection task, and malware family labels for the attribution task). We consider two primary sources for such reports based on the collection environment. First, we collect the reports from a software sandbox

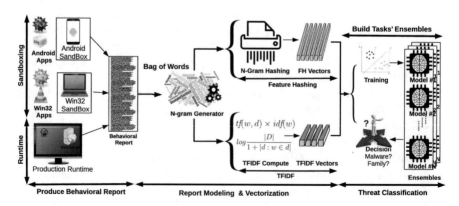

Fig. 5.3 MalDy methodology overview

environment [4], in which we execute the binary program, malware, or benign, in a controlled environment (mostly virtual machines). The primary usage of sandboxing in security investigation is to check and analyze the maliciousness of programs. Second, we could collect behavioral reports from a production system in the form of system logs of the running apps. The goal is to investigate the sanity of the apps during their execution. As presented in Sect. 5.2, MalDy employs a word-based approach to model behavioral reports.

5.3.2 Report Vectorization

In this section, we answer the question: how can we model the words in the behavioral report to fit in our classification component? Previous solutions [5, 6] select specific features from the behavioral reports by: (1) extracting relevant security features and (2) manually inspecting and selecting from these features [5]. This process requires manual intervention of the security analyst. Also, it is not scalable since he/she needs to redo this process manually for each new type of behavioral report. In contrast, we are looking for features (words in our case) representation that allows for an automatic feature engineering without the intervention of a security expert.

5.3.3 Build Models

MalDy framework utilizes a supervised machine learning technique to build its malware investigation models. In this respect, MalDy is composed of a set of models, and each model has a specific purpose. First, we have the threat detection model that finds out the maliciousness likelihood of a given app from its behavioral report. Afterward, the remaining machine learning models aim to investigate individual family threats separately. MalDy uses a model for each possible threat. In our case, we have a malware detection model along with a set of malware family attribution models. In this phase, we build each model separately using X_{build}. All the models are employing a binary classification to quantify the likelihood of a specific threat. In the process of building MalDy models, we evaluate different classification algorithms to compare their performance. Furthermore, we tune each ML algorithm classification performance under an array of hyper-parameters (different for each ML algorithm). The tuning is a completely automatic process; the investigator only needs to provide X_{build}. We train each investigation model on X_{train} and tune the models performance on X_{valid} by finding the best hyper-parameters as presented in Algorithm 5. Afterward, we determine the optimum decision thresholds for each model using its performance on X_{valid}. At the end of this stage, we have a list of optimum models' tuples $Opt = \{< c_0, th_0, params_0 >, < c_1, th_1, params_1 > , .., < c_c, th_c, params_c >\}$, where c is the number of explored classification algorithms. A tuple $< c_i, th_i, params_i >$ defines the optimum hyper-parameters $params_i$ and decision threshold th_i for ML classification algorithm c_i.

Algorithm 5: Build models algorithm

Input: X_{build}: build set
Output: Opt: optimum models' tuples
$X_{train}, X_{valid} = X_{build}$
for c in MLAlgorithms **do**
$\quad$ score = 0 **for** $params$ in c.params_array **do**
$\quad\quad$ model = train(alg, X_{train}, $params$) ;
$\quad\quad$ s, th = validate(model, X_{valid}) ;
$\quad\quad$ **if** $s > score$ **then**
$\quad\quad\quad$ ct = $< c, th, params >$;
$\quad\quad$ **end**
$\quad$ **end**
$\quad$ Opt.add(ct)
end
return Opt

5.3.4 Ensemble Composition

Previously, we discuss the process of building and tuning individual classification models for specific investigation tasks (malware detection, family one threat attribution, family two threat attribution, etc.). In this phase, we construct an ensemble model from a set of models generated using the optimum parameters computed previously (Sect. 5.3.3), such that the ensemble outperforms any underlying model. We take each set of optimally trained models $\{(C_1, th_1), (C_2, th_2), .., (C_h, th_h)\}$ for a specific threat investigation task and unify them into an ensemble E. The latter utilizes the weighted majority-voting mechanism across the individual model's outcomes for a specific investigation task. Equation 5.1 shows the computation of the final outcome for one ensemble E, where w_i is the weight given for a single model. The current implementation employs equal weights for the ensemble's models. This phase produces **MalDy** ensembles, $\{E^1_{Detection}, E^2_{Family1}, E^3_{Family2} .., E^T_{familyJ}\}$, a malware detection ensemble, and an ensemble for each malware family.

$$\hat{y} = E(x) = sign \left(\sum_i^{|E|} w_i C_i(x, th_i) \right)$$

$$= \begin{cases} +1 : \sum_i (w_i C_i) \geq 0 \\ -1 : \sum_i (w_i C_i) < 0 \end{cases}$$

(5.1)

5.3.5   Ensemble Prediction Process

MalDy prediction process is divided into two phases, as depicted in Algorithm 6. First, given a behavioral report, we generate the feature vector x using TF-IDF or FH vectorization techniques. Afterward, the detection ensemble $E_{detection}$ checks the maliciousness likelihood of the feature vector x. If the maliciousness detection is positive, we proceed to the family threat attribution. Since the family threat ensembles, $\{E^2_{Family1}, E^3_{Family2} ..., E^T_{familyJ}\}$, are independent, we compute the outcomes of each family ensemble E_{family_i}. MalDy flags a malware family threat if and only if the majority voting is above a given voting threshold vth (computed using X_{valid}). In the case where no family threat is flagged by the family ensembles, MalDy tags the current sample as an unknown threat. Also, in the case of multiple flagged families, MalDy selects the family with the highest probability and provides the security investigator with sorted flagged families according to the corresponding likelihood probabilities. The separation between the family attribution models makes MalDy more flexible to update. Adding a new family threat will need only to train, tune, and calibrate the family model without affecting the rest of the framework ensembles.

Algorithm 6: Prediction algorithm

Input: $report$: Report
Output: D: Decision
$E_{detection} = E^1_{Detection}$;
$E_{family} = \{E^2_{Family1}, ..., E^T_{familyJ}\}$;
$x = \text{Vectorize}(report)$;
detection_result = $E_{detection}(x)$;
if $detection\_result < 0$ **then**
 | **return** detection_result ;
end
for E_{F_i} in E_{family} **do**
 | family_result = $E_{F_i}(x)$;
end
return detection_result, family_result ;

5.4   MalDy Framework

In this section, we present the essential techniques used in MalDy framework, namely, N-grams [7], feature hashing (FH), and term frequency-inverse document frequency (TFIDF). Furthermore, we present the explored and tuned machine learning algorithms during the model building phase (Sect. 5.4.1).

We describe the components of the **MalDy** related to the automatic security feature engineering process. We compute word N-grams on behavioral reports by counting the word sequences of size N. Notice that N-grams are extracted using a moving forward window (of size N) by one step and incrementing the counter of the found feature (word sequence in the window) by one. The window size N is a hyper-parameter in the **MalDy** framework. N-gram computation happens simultaneously with the vectorization using FH or TFIDF in the form of a pipeline to prevent computation and memory issues due to the high dimensionality of N-grams. From a security perspective, the N-grams tool can produce distinguishable features between the different variations of an event log compared to single word (1-grams) features. The performance of the malware investigation is profoundly affected by the features generated using N-grams (where $N > 0$). Based on the BoW model, **MalDy** considers the count of unique N-grams as features that will be leveraged through a pipeline to FH or TFIDF as presented in the previous sections.

5.4.1 Machine Learning Algorithms

Table 5.1 shows the candidate machine learning classification algorithms for **MalDy** framework. The candidates represent the most used classification algorithms and come from different learning categories, such as decision tree-based learning algorithms. Also, all these algorithms have efficient public implementations. We chose to exclude logistic regression from the list due to the superiority of SVM in most cases. KNN may consume a lot of memory resources during production because it needs all the training dataset to be deployed in the production environment. However, we keep KNN in the **MalDy** candidate list because of its unique fast update feature. Updating KNN in a production environment requires only to update the train dataset, and we do not need to retrain the model. This option could be beneficial in certain malware analysis cases.

Table 5.1 Explored machine learning classifiers

Classifier category	Classifier algorithm	Chosen
Tree	CART	✓
	Random forest	✓
	Extremely randomized trees	✓
General	K-nearest neighbor (KNN)	✓
	Support vector machine (SVM)	✓
	Logistic regression	✗
	XGBoost	✓

5.5   Evaluation Results

5.5.1   Evaluation Datasets

Table 5.2 presents different datasets used to evaluate MalDy framework. We focus on Android and Win32 platforms to prove the portability of MalDy. All the used datasets are publicly available except Win32 Malware dataset, which is provided by a third-party security vendor. Behavioral reports are generated using DroidBox [1] and ThreatAnalayzer[2] for Android and Win32, respectively.

Next, we evaluate MalDy on different datasets and various settings. Specifically, we assess the effectiveness of the word-based approach for malware detection and family attribution on Android malware behavior reports. We evaluate the portability and the MalDy concept on other platforms (e.g., Win32 malware) behavioral reports. Finally, we measure the efficiency of MalDy under different machine learning classifiers and vectorization techniques. During the evaluation, we answer some other questions related to the comparison between the vectorization techniques, and the used classifiers in terms of effectiveness and efficiency. Also, we show the effect of the training sets size and the usage of machine learning ensemble technique on the framework performance.

5.5.2   Effectiveness

Figure 5.4 shows the detection and the attribution performance under various settings and datasets. The settings refer to the used classifiers in ML ensembles and their hyper-parameters, as shown in Table 5.4. Figure 5.4a depicts the overall performance of MalDy. In the detection, MalDy achieves 90% F1-score (100% maximum and about 80% minimum) on average under the various settings (classifi-

Table 5.2 Evaluation datasets (D: DroidBox, T: ThreatAnalyzer)

Platform	Dataset	Sandbox	Tag	#Sample/#Family
Android	MalGenome [8]	D	Malware	1k/10
	Drebin [9]	D	Malware	5k/10
	Maldozer [10]	D	Malware	20k/20
	AndroZoo [11]	D	Benign	15k/-
	PlayDrone[3]	D	Benign	15k/-
Win32	Malware[4]	T	Malware	20k/15

---

[2]threattrack.com.

[3]https://archive.org/details/android_apps.

[4]https://threattrack.com/.

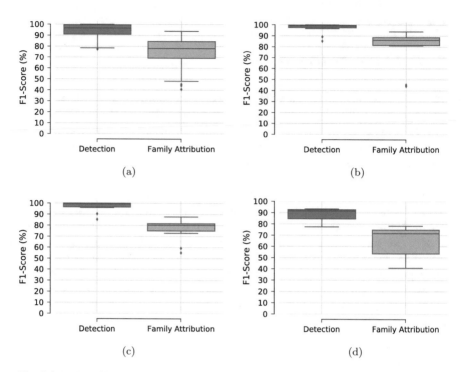

Fig. 5.4 MalDy effectiveness performance. (**a**) General. (**b**) Malgenome. (**c**) Drebin. (**d**) Maldozer

cation models, vectorization techniques, hyper-parameters tuning, single model, and models ensemble). On the other hand, in the family attribution task, MalDy shows over 80% F1-score (family attribution is a harder task than the detection task) in various settings. More granular results for each dataset are shown in Fig. 5.4b, c, and d for Malgenome [8], Drebin [9], and Maldozer[10] datasets, respectively. Notice that Fig. 5.4a combines the performance of baseline (worst performance), tuned, and ensemble models and summarizes the results in Table 5.3.

5.5.2.1 Classifier Effect

The results in Fig. 5.5, Tables 5.3, and 5.4 confirm the effectiveness of MalDy framework and its word-based approach. Figure 5.5 presents the effectiveness performance of MalDy using the different classifiers for the final ensemble models. Figure 5.5a shows the combined performance of the detection and family attribution. All the ensembles achieved a good F1-score, and XGBoost ensemble shows the highest scores. Figure 5.5b confirms the previous scores for the detection task. Also, Fig. 5.5c presents the malware family attribution scores per ML classifier. More details on classifiers performance are depicted in Table 5.4.

Table 5.3 Tuning effect on performance

	Detection (F1 %)			Attribution (F1 %)		
	Base	Tuned	Ens	Base	Tuned	Ens
General						
Mean	86.06	90.47	94.21	63.42	67.91	73.82
Std	6.67	6.71	6.53	15.94	15.92	14.68
Min	69.56	73.63	77.48	30.14	34.76	40.75
25%	83.58	88.14	90.97	50.90	55.58	69.07
50%	85.29	89.62	96.63	68.81	73.31	78.21
75%	91.94	96.50	99.58	73.60	78.07	84.52
Max	92.81	97.63	100.0	86.09	90.41	93.78
Genome						
Mean	88.78	93.23	97.06	71.19	75.67	79.92
Std	5.26	5.46	4.80	16.66	16.76	16.81
Min	77.46	81.69	85.23	36.10	40.10	44.09
25%	85.21	89.48	97.43	72.36	77.03	81.47
50%	91.82	96.29	99.04	76.66	81.46	86.16
75%	92.13	96.68	99.71	80.72	84.82	88.61
Max	92.81	97.63	100.0	86.09	90.41	93.78
Drebin						
Mean	88.92	93.34	97.18	65.97	70.37	76.47
Std	4.93	4.83	4.65	9.23	9.14	9.82
Min	78.36	83.35	85.37	47.75	52.40	55.10
25%	84.95	89.34	96.56	61.67	65.88	75.05
50%	91.60	95.86	99.47	69.62	74.30	80.16
75%	92.25	96.53	100.0	72.68	76.91	81.61
Max	92.78	97.55	100.0	76.28	80.54	87.71
Maldozer						
mean	80.48	84.85	88.38	53.11	57.68	65.06
std	6.22	6.20	5.95	16.03	15.99	13.22
min	69.56	73.63	77.48	30.14	34.76	40.75
25%	75.69	80.13	84.56	39.27	43.43	53.65
50%	84.20	88.68	91.58	56.62	61.03	71.65
75%	84.88	89.01	92.72	67.34	71.89	74.78
max	85.68	89.97	93.39	71.17	76.04	78.30

5.5.2.2 Effect of the Vectorization Technique

Figure 5.6 shows the effectof vectorization techniques on the detection and the attribution performance. Figure 5.6a depicts the overall combined performance under various settings. As depicted in Fig. 5.6a, Feature hashing and TF-IDF show a very similar performance. In the detection task, the vectorization techniques' F1-scores are almost identical to those presented in Fig. 5.6b. We notice a higher overall attribution scoreusing TF-IDF compared to FH, as shown in Fig. 5.6c. However,

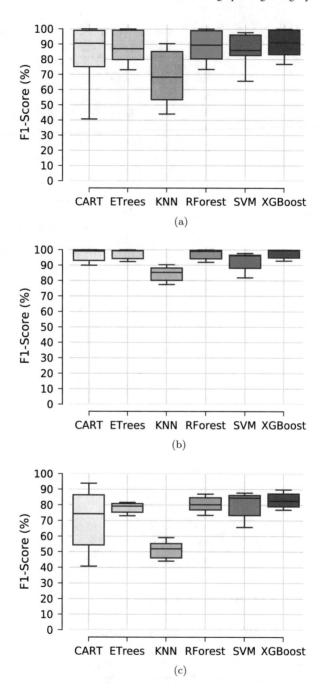

Fig. 5.5 Effectiveness per machine learning classifier. (**a**) General. (**b**) Detection. (**c**) Attribution

Table 5.4 Android Malware detection

Settings			Attribution F1-score (%)			Detection F1-score (%)			
Model	Dataset	Vector	Base	Tuned	Ensemble	Base	Tuned	Ensemble	FPR(%)
CART	Drebin	Hashing	64.93	68.94	72.92	91.55	95.70	99.40	00.64
	Drebin	tfidf	68.12	72.48	75.76	92.48	96.97	100.0	00.00
	Genome	Hashing	82.59	87.28	89.90	91.79	96.70	98.88	00.68
	Genome	tfidf	86.09	90.41	93.78	92.25	96.50	100.0	00.00
	Maldozer	Hashing	33.65	38.56	40.75	82.59	87.18	90.00	06.92
	Maldozer	tfidf	40.14	44.21	48.07	83.92	88.67	91.16	04.91
ETrees	Drebin	Hashing	72.84	77.27	80.41	91.65	95.77	99.54	00.23
	Drebin	tfidf	71.12	76.12	78.13	92.78	97.55	100.0	00.00
	Genome	Hashing	74.41	79.20	81.63	91.91	96.68	99.14	00.16
	Genome	tfidf	73.83	78.65	81.02	92.09	96.61	99.57	00.03
	Maldozer	Hashing	65.23	69.34	73.13	84.56	88.70	92.42	06.53
	Maldozer	tfidf	67.14	71.85	74.42	84.84	88.94	92.74	06.41
KNN	Drebin	Hashing	47.75	52.40	55.10	78.36	83.35	85.37	12.86
	Drebin	tfidf	51.87	56.53	59.20	82.48	86.57	90.40	05.83
	Genome	Hashing	36.10	40.10	44.09	77.46	81.69	85.23	07.01
	Genome	tfidf	37.66	42.01	45.31	81.22	85.30	89.13	02.10
	Maldozer	Hashing	41.68	46.67	48.69	69.56	73.63	77.48	26.21
	Maldozer	tfidf	48.02	52.73	55.31	70.94	75.36	78.51	03.86
RForest	Drebin	Hashing	72.63	76.80	80.46	91.54	95.95	99.12	00.99
	Drebin	tfidf	72.15	76.40	79.91	92.31	96.62	100.0	00.00
	Genome	Hashing	78.92	83.73	86.12	91.37	95.79	98.95	00.68
	Genome	tfidf	79.45	83.90	87.00	92.75	97.49	100.0	00.00
	Maldozer	Hashing	66.06	70.72	73.41	84.49	88.96	92.01	07.37
	Maldozer	tfidf	67.96	72.04	75.89	85.07	89.41	92.72	06.10
SVM	Drebin	Hashing	57.35	61.95	82.92	84.50	89.33	96.08	00.86
	Drebin	tfidf	63.11	67.19	87.71	85.11	89.35	96.73	01.15
	Genome	Hashing	69.99	74.68	86.08	85.47	89.83	96.54	00.19
	Genome	tfidf	73.16	77.82	86.20	84.46	88.46	97.73	00.39
	Maldozer	Hashing	30.14	34.76	65.76	72.32	77.12	81.88	15.82
	Maldozer	tfidf	36.69	41.09	70.18	76.82	81.14	85.46	08.56
XGBoost	Drebin	Hashing	76.28	80.54	**84.01**	92.05	96.50	**99.61**	**00.29**
	Drebin	tfidf	73.53	77.88	**81.18**	92.23	96.45	**100.0**	**00.00**
	Genome	Hashing	81.80	85.84	**89.75**	91.86	96.09	**99.62**	**00.32**
	Genome	tfidf	80.36	84.48	**88.24**	92.81	97.63	**100.0**	**00.00**
	Maldozer	Hashing	71.17	76.04	**78.30**	85.68	89.97	**93.39**	**05.86**
	Maldozer	tfidf	69.51	74.15	**76.87**	85.01	89.16	**92.86**	**06.05**

The bold values represent the values of XGBoost that are the best compared to other techniques

Fig. 5.6 Effectiveness per
vectorization technique. (**a**)
General. (**b**) Detection. (**c**)
Attribution

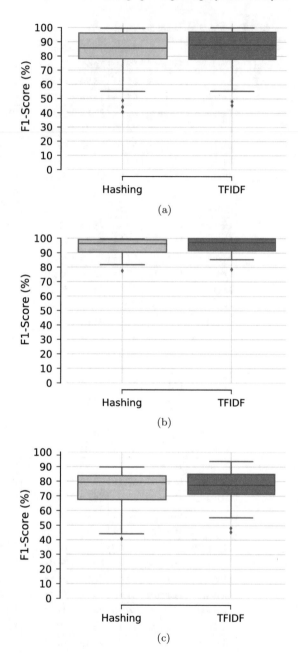

there are some cases where FH outperforms TF-IDF. For instance, XGBoost
achieves a higher attribution score under feature hashing vectorization, as shown
in Table 5.4.

5.5.2.3   Effect of Tuning Hyper-Parameters

Figure 5.7 illustrates the effects of tuning and ensemble phases on the overall performance of MalDy. In the detection task, as in Fig. 5.7a, the ensemble improves the performance by 10% (F1-score) over the base model. The ensemble is composed of a set of tuned models that already outperform the base model. In the attribution task, the ensemble improves the F1-score by 9%, as shown in Fig. 5.7b.

5.5.3   Portability

In the following, we assess the portability of the MalDy by applying the framework on a new type of behavioral reports. Also, we investigate the appropriate training dataset size for MalDy to achieve a good result. We report only the results of the attribution task on Win32 malware because currently we do not have a dataset of Win32 benign behavioral reports for the detection task.

Fig. 5.7 Ensemble performance and tuning effect. (**a**) Detection. (**b**) Attribution

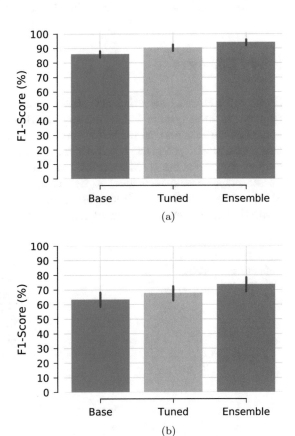

5.5.3.1 MalDy on Win32 Malware

Table 5.5 presents MalDy attribution performance in terms of F1-score. In contrast with previous results, we train MalDy models on only 2k (10%) out of 20k report' dataset (Table 5.2). The rest of the reports are used for testing (18k reports, or 80%). Despite that, MalDy achieves high scores that reach 95%. The results in Table 5.5 illustrate the portability of MalDy, which increases the utility of the framework across the different platforms and environments.

5.5.3.2 MalDy Train Dataset Size

Using the Win32 malware dataset (Table 5.5), we investigate the training set size hyper-parameter for MalDy to achieve good results. Figure 5.8 illustrates the outcome of our analysis for both vectorization techniques and the different classifiers. We notice the high scores of MalDy even with relatively small datasets. This is made very clear in the case where MalDy uses the SVM ensemble, which corresponds to a 87% F1-score with only 200 training samples.

5.5.4 Efficiency

Figure 5.9 illustrates the efficiency of MalDy by showing the average runtime required to investigate a behavioral report. The runtime is composed of the preprocessing time and the prediction time. As depicted in Fig. 5.9, MalDy needs only about 0.03 s per given report for all the ensembles and the preprocessing settings except for the SVM ensemble. The latter requires 0.2–0.5 s (depending on the preprocessing technique) to decide about a given report. Although the SVM ensemble needs a small training set to achieve good results, it is costly in production environment in terms of runtime. Therefore, the security investigator

Table 5.5 System performance on Win32 Malware

	Ensemble F1-score(%)	
Model	Hashing	TFIDF
CART	82.35	82.74
ETrees	92.62	92.67
KNN	76.48	80.90
RForest	91.90	92.74
SVM	91.97	91.26
XGBoost	**94.86**	**95.43**

The bold values represent the values of XGBoost that are the best compared to other techniques

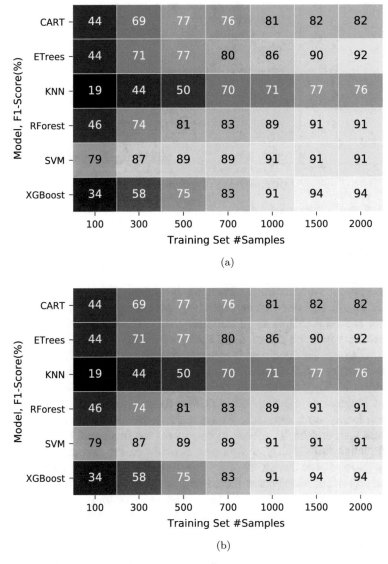

Fig. 5.8 Win32 performance and effect of the training size. (**a**) Hashing (F1-Score %). (**b**) TF-IDF (F1-Score %)

could customize the MalDy to suit particular analysis priorities. The efficiency experiments have been conducted on Intel(R) Xeon(R) CPU E52630 (128G RAM), using only one CPU core.

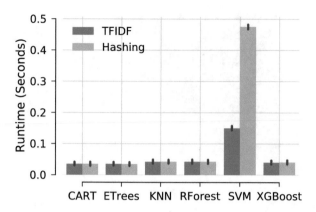

Fig. 5.9 Overall framework efficiency

5.6   Summary

In this chapter, we detailed a pioneering investigation on the use of dynamic features for Android malware fingerprinting. We leverage state-of-the-art *machine learning* and NLP techniques to elaborate a portable, effective, and yet efficient framework for malware detection and family attribution. The key concept involves the modeling of behavioral reports using the bag of words model. Furthermore, we leverage advanced NLP and ML techniques to build discriminative machine learning ensembles. MalDy achieves over 94% F1-score in Android malware detection task on Malgenome, Drebin, and MalDozer datasets and more than 90% in the malware family attribution task. We demonstrate MalDy portability by applying the framework on Win32 malware reports where the framework achieves 94% on the attribution task. MalDy performance depends on the execution environment reporting system, and the quality of the reporting affects its performance.

In the previous chapter and the current one, we focus on Android malware detection using machine learning classification and clustering, respectively, based on dynamic and static analyses features. In the next chapter, we propose a system for the mining of cyber-threat networks, which are composed of malicious IP addresses and domain names, starting from the detected Android malware. The aforementioned cyber-threat networks support malicious apps and act as a backend service. Thus, finding malicious cyber-threat networks represents a natural progression following the detection of Android malware as proposed in previous chapters (Chaps. 5 and 4).

References

1. DroidBox, https://github.com/pjlantz/droidboxh (2021). Accessed May 2016
2. Android Emulator, https://developer.android.com/studio/run/emulator (2016). Accessed on February 2016
3. Android Software Developer Kit (SDK), https://developer.android.com/studio/ (2019). Accessed June 2019
4. C. Willems, T. Holz, F.C. Freiling, Toward automated dynamic malware analysis using CWSandbox. IEEE Secur. Privacy **5**(2), 32–39 (2007)
5. S. Chen, M. Xue, Z. Tang, L. Xu, H. Zhu, Stormdroid: A streaminglized machine learning-based system for detecting android malware, in *Proceedings of the 11th ACM on Asia Conference on Computer and Communications Security, AsiaCCS 2016, Xi'an, May 30–June 3, 2016* (2016), pp. 377–388
6. K. Rieck, P. Trinius, C. Willems, T. Holz, Automatic analysis of malware behavior using machine learning. J. Comput. Secur. **19**(4), 639–668 (2011)
7. T. Abou-Assaleh, N. Cercone, V. Keselj, R. Sweidan, N-gram-based detection of new malicious code, in *28th International Computer Software and Applications Conference (COMPSAC 2004), Design and Assessment of Trustworthy Software-Based Systems, 27–30 September 2004, Hong Kong, Workshop Papers* (2004), pp. 41–42
8. Y. Zhou, X. Jiang, Dissecting android malware: Characterization and evolution, in *IEEE Symposium on Security and Privacy, SP 2012, 21–23 May 2012, San Francisco,* , pp. 95–109 (2012)
9. D. Arp, M. Spreitzenbarth, M. Hubner, H. Gascon, and K. Rieck, DREBIN: effective and explainable detection of android malware in your pocket, in *21st Annual Network and Distributed System Security Symposium, NDSS 2014, San Diego, California, February 23–26, 2014* (2014)
10. E.B. Karbab, M. Debbabi, A. Derhab, D. Mouheb, Maldozer: Automatic framework for android malware detection using deep learning. Digit. Investig. **24**, S48–S59 (2018)
11. K. Allix, T.F. Bissyandé, J. Klein, Y.L. Traon, Androzoo: collecting millions of android apps for the research community, in *Proceedings of the 13th International Conference on Mining Software Repositories, MSR 2016, Austin, May 14–22, 2016* (2016), pp. 468–471
12. A. Kharraz, S. Arshad, C. Mulliner, W.K. Robertson, E. Kirda, UNVEIL: A large-scale, automated approach to detecting ransomware, in *25th USENIX Security Symposium, USENIX Security 16, Austin, August 10–12, 2016* (2016), pp. 757–772
13. D. Sgandurra, L. Muñoz-González, R. Mohsen, E.C. Lupu, Automated dynamic analysis of ransomware: Benefits, limitations and use for detection. CoRR **abs/1609.03020** (2016).
14. C. Kolbitsch, P.M. Comparetti, C. Kruegel, E. Kirda, X. Zhou, X. Wang, Effective and efficient malware detection at the end host, in *USENIX Security Symposium* (2009), pp. 351–366
15. M.K. Alzaylaee, S.Y. Yerima, S. Sezer, Dynalog: An automated dynamic analysis framework for characterizing android applications. CoRR **abs/1607.08166** (2016).
16. L. Chen, M. Zhang, C. Yang, R. Sahita, POSTER: semi-supervised classification for dynamic android malware detection, in *Proceedings of the 2017 ACM SIGSAC Conference on Computer and Communications Security, CCS 2017, Dallas, October 30–November 03, 2017* (2017), pp. 2479–2481

Chapter 6
Fingerprinting Cyber-Infrastructures of Android Malware

In this chapter, we propose ToGather, an automatic investigation framework for Android malware cyber-infrastructures. In our context, a malware cyber-infrastructure is a set of IP addresses and domain names orchestrated together to serve as a backend for malicious activities, including malicious apps. ToGather framework is a set of techniques and tools together with security feeds, which automatically build a situational awareness about Android malware cyber-infrastructures. ToGather characterizes the cyber-infrastructure starting from Android malware samples to relate the malware to the corresponding network footprint in terms of IPs and domains. ToGather goes even a step further by dividing this cyber-infrastructure into sub-infrastructure components based on the connectivity between nodes. The result is in the segmentation of the global threat network into multiple network communities representing many granular sub-cyber-infrastructures. To this end, ToGather leverages cyber-threat intelligence that is derived from various sources such as spam, Windows malware, darknet, and passive DNS to ascribe cyber-threats to the corresponding cyber-infrastructure. Accordingly, the input of ToGather framework is made of malware samples together with security feeds, while the output represents networks of cyber-infrastructures together with their network footprint, which provides the security practitioner an overview of Android malware cyber-activities on the Internet.

6.1 Threat Model

We position ToGather as a detector of malicious cyber-infrastructures of Android malware. It is designed to uncover threat networks and sub-networks starting from Android malware samples. ToGather does not guarantee zero false positives due to the large number of benign domain names and IP addresses that might not be filtered out using ToGather whitelists. ToGather is resilient to obfuscation during

E. B. Karbab et al., *Android Malware Detection Using Machine Learning*, Advances in Information Security 86, https://doi.org/10.1007/978-3-030-74664-3_6

the extraction of network information from Android malware because it applies both static and dynamic analyses. Hence, if the static content is heavily obfuscated, ToGather is still able to collect IP addresses and domain names from dynamic analysis reports.

6.2 Usage Scenarios

ToGather is designed to be practical and efficient in the hands of security practitioners. (1) Security analysts might use ToGather framework as an investigation tool to minimize the efforts of generating threat networks for a given Android malware family. The analyst leverages the IP addresses and domain names ordered by their importance in the generated threat network to prioritize the takedown and mitigation operations. (2) ToGather acts as a monitoring system. It analyzes a feed of Android malware (e.g., new samples daily) to generate a snapshot of the threat network and thus uncover malicious activities (e.g., spamming and phishing). Periodic reporting gives insights into the evolution and the malicious behaviors of a given malware family over time.

6.3 Methodology

In this section, we present the overall workflow of ToGather framework, as shown in Fig. 6.1, starting from Android malware samples and ending with the produced relevant threat intelligence:

(1) The first step in ToGather consists of deriving network information from Android samples in a given analysis window (e.g., day, week, month) whether the samples are from the same malware family or not. However, we consider one malware family as a typical use case of ToGather, as presented in the evaluation section. ToGather conducts dynamic and static analyses where each analysis produces a report for each Android malware sample. Therefore, we produce dynamic and static analyses reports for each malware sample. Leveraging both analysis types enhances the resiliency of ToGather against common obfus-

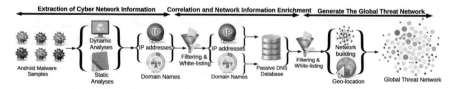

Fig. 6.1 ToGather approach overview

cation techniques, which hide relevant information about malicious activities such as domain names and IP addresses (network information). Afterward, ToGather extracts network information (IP addresses and domain names) by parsing the related text blocks (strings) from analysis reports and applies a simple text pattern search. In static analysis, we mainly concentrate on the Dalvik compiled code (classes.dex) for such extraction. We collect network information more efficiently from dynamic analysis reports since they are more structured and have labeled fields.

(2) Next, we filter the extracted network identifiers from noise information such as non-routed IP addresses. Also, we filter domain names and URLs that use Unicode characters. For the current ToGather implementation, we consider domain names and URLs written only in the standard English/Latin alphabet. In the case of URL links, we keep only domains. To this end, we have a set of valid IP addresses and domain names found in Android malware. It is important to notice that we keep malware hashes related to network information (IPs and domains) during all the workflow steps of ToGather. To minimize false positives, ToGather applies whitelisting mechanisms. For domain names, ToGather leverages Alexa [1] and Quantcast [2] (more than one million domain names). However, the number of white domain names is a hyper-parameter of ToGather that can be used to control the number of false positives. In the case of IP addresses, we leverage a set of public white IPs such as Google DNS servers and other ones [3]. It is important to emphasize that ToGather considers public cloud vendor IPs and domain names as a whitelist. The aim is to observe and then gain insight into the use of the cloud infrastructure by Android malware. This idea originates from the observation that Android malicious apps (and malware in general) make more use of the cloud as a low-cost infrastructure for their malicious activity.

(3) In this step, we propose a mechanism to enhance and enrich the malicious network information to cover related domains and IPs. In essence, ToGather aims at answering the following questions: (1) What are the IP addresses of current malicious domains? Here we investigate the IP addresses of server machines that host malicious activities that are most likely related to the analyzed Android malware. (2) What are the domain names pointing to the current malicious IP addresses? The intuition is that a malicious server machine with a given IP address could host various malicious contents, and the adversary could use multiple domains pointing to such contents. To answer this question, ToGather has a module to enrich network information using passive DNS replication. The latter is a technology that builds zone replicas without the cooperation from zone administrators, based on captured name server responses, as presented in Sect. 6.3.3.1. We use the network information, whether IP addresses or domains, as parameters of two functions applied on a passive DNS database. The goal of the function is to enrich the list of domains and IP addresses that could be part of the adversary threat network. The enrichment services are: (1) GetIP(Domain): This function takes a domain as a parameter to query the passive DNS database. The result is all IP addresses pointing to the domain. (2) GetDomain(IP): This function gets all the domains that resolve to the IP address given as a parameter.

We consider passive DNS correlation for two reasons: (1) A small number of Android malware samples generally yields limited network information. (2) Security practitioners aim at having a more in-depth situational awareness about malware Internet activity. As such, they would like to consider all related IPs and domain names. The result of the correlation is a set of IP addresses and domain names inferred using passive DNS related to Android malware apps . The correlation results could, however, overwhelm the investigation process. Passive DNS correlation is therefore optional if we have a significant number of samples from a given Android family. The correlation with passive DNS could produce some known benign entries. For this reason, we filter the likely harmless network information by matching the newly found IP addresses against top Alexa [1] and Quantcast [2] domain names and known public IP addresses [4].

(4) At this stage, we have a set of network information tagged by malware hashes. To extract relevant and actionable intelligence, ToGather aggregates all the previous records into a heterogeneous network with different types of nodes: *malware hashes*, *IP addresses*, and *domain names*. We consider the heterogeneous network that is extracted from a given Android malware family as the malicious activity map of that family on the Internet. We call such a heterogeneous network a *threat network*. Furthermore, ToGather produces homogenous networks by executing multiple projections according to the node type (IP address or domain name). Therefore, ToGather produces three homogeneous graphs, one only considers IP addresses connections, the other only considers domain name connections, and a threat network with IPs and domains as network information. The Graph homogeneity is required to apply graph partitioning on domain threat network and network information threat network.

(5) Further, ToGather aims at producing more granular graphs (see Fig. 6.2) from the generated threat networks derived in the previous step. In this respect, ToGather checks the possibility of community identification in these

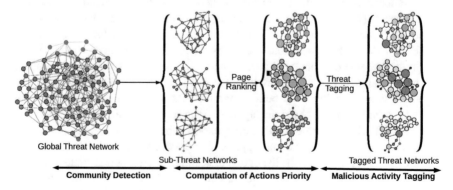

Fig. 6.2 Graph analysis overview

threat networks based on the connectivity between nodes. The higher is the connectivity between the nodes in a particular area of the network, the more is the possibility to have a malicious community. For community detection (Sect. 6.3.1), we adopt a highly scalable algorithm [5] to enhance ToGather community detection module. The intuition behind using the community concept is as follows: (1) Considering ToGather typical usage scenario, where we enter Android malicious apps from the same family, the community could define different threat networks that are related to the malicious activities. In other words, either one adversary is using these threat networks as backups, or we have multiple adversaries instead. In the case of Android malware, the second hypothesis is more plausible because of the low cost of repackaging of existing malware samples to suit the need of the adversary. (2) In case ToGather receives Android malware from different families, the communities are interpreted as the threat networks of different Android malware families to focus on the relation between them. The output of this step is a set of threat networks related to IPs, domains, as well as network information and their communities (sub-threat networks).

(6) To produce actionable cyber-threat intelligence, we leverage the page ranking algorithm (Sect. 6.3.2) to deliver ranking scores for critical nodes of a given (sub)-threat network. Consequently, the investigator should have some priority list when it comes to mitigation or takedown of nodes that are associated with a malicious cyber-infrastructure. As a result, ToGather produces a threat network for each Android malware family together with the ranking of each node. Because ToGather generates multiple homogeneous graphs based on the node type (IP, domain, network information), it produces different ranking lists. Therefore, the security practitioner has the opportunity of selecting the node type during the mitigation or the takedown to protect his system. Also, it is essential to mention that it is expensive for the adversary to get new IP addresses. In contrast, domain names could be frequently changed due to their affordability.

(7) We do not focus only on Android malware. Instead, we aim to gain insights into the shared network IP and domains of the analyzed Android malware samples with other platform malware families. Indeed, an adversary could have many malicious activities in several operating systems to achieve wider coverage. Therefore, similarly to the first step, we conduct dynamic and static analyses on Windows and Linux malware samples to extract the corresponding network information. The same step is applied to this network information. Afterward, we correlate the Android network information with the non-Android malware information to discover another dimension of the adversary network. The result will be all IP addresses and domains of Android malware in addition to all network records of a given non-Android malware family if they share some network information. It is essential to notice that malware families also label information networks of non-Android malware.

(8) In this final workflow step of ToGather, we leverage other intelligence sources to label malicious activities that are committed by the discovered

threat networks. The current ToGather implementation includes the correlation with spam emails, reconnaissance traces, and phishing URLs. We consider ToGather as an active service that receives at every epoch time (day, week, month) Android malware with the corresponding family (the typical use case) and produces valuable intelligence about this malware family.

6.3.1 Threat Communities Detection

A scalable community detection algorithm is essential to extract communities from the threat network. For this reason, we empower ToGather with the Fast Unfolding Community Detection algorithm [5], which can scale to billions of network links. The algorithm achieves excellent results by measuring the *modularity* of communities. The latter is a scalar value $M \in [-1, 1]$ that measures the density of edges inside a given community compared to the edges between communities. The algorithm uses an approximation of the modularity since finding the exact value is computationally hard [5]. The main reason to choose the algorithm proposed in [5] is its scalability. As depicted in Fig. 6.3, we apply the community detection on a million-node graph with a medium density ($P = 0.001$ probability of connecting a node A is another node B in the generated network), which we believe has a similar density to the threat network generated from Android malware samples. For the sake of completeness, we perform the same experiment on graphs with a different probability P. As presented in Fig. 6.4c, we can detect communities in 30,000-node graphs with ultra high density (unrealistic) in a relatively small (compared to the time dedicated to the investigation) amount of time (3 h).

The previous algorithm requires as input a homogeneous network to work correctly. In our case, the threat network generated from the network information is heterogeneous because it contains two main node types: (1) the malware sample identifier, which is the cryptographic hash of the malware sample; (2) the network

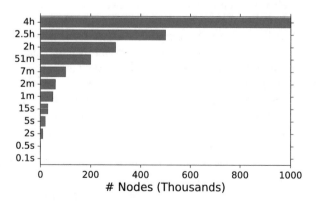

Fig. 6.3 Scalability of the community detection

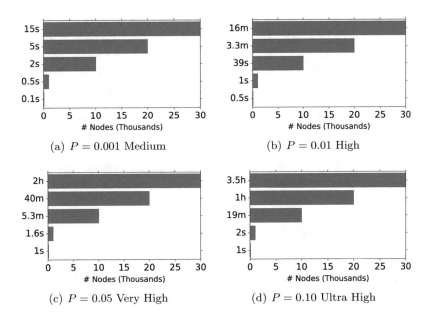

Fig. 6.4 Graph density versus scalability. (**a**) $P = 0.001$ medium. (**b**) $P = 0.01$ high. (**c**) $P = 0.05$ very high. (**d**) $P = 0.10$ ultra high

information: the domain names and the IPv4 addresses. Also, we apply the projection on the first heterogeneous network to generate homogeneous graphs. To do so, ToGather makes the graph projection by abstracting away from malware identifier while keeping network information, i.e., if malware connects to two IPs, the projection would produce only the two IPs involved in the connection. To this end, we get different projection results based on the node abstraction: (a) general threat network containing both IP addresses and domain names, (b) IP threat network containing only IP addresses, and (c) domains related threat network containing only domain names.

6.3.2 Action Prioritization

From community detection, ToGather checks if there are possible sub-graphs in the threat networks based on node connectivity. Even though threat networks zoom into malicious cyber-infrastructures of a given Android malware family, it is difficult for the security practitioner to tackle the whole threat network at once. For this reason, ToGather proposes an action priority system. The latter takes the IP, domain (or both), and the threat network and produces an action priority list based on the maliciousness of each node. By leveraging the graph structure of the threat network, we measure the maliciousness of a given node by its degree, meaning the number of edges that connect it to other nodes. From a security point of view, the more connections an IP or domain has, the more it is important for a malicious cyber-infrastructure. Therefore, we build a priority list sorted by the importance, that an IP, or a domain can inflict in terms of malicious activity. The importance of a given node in a network graph is known as *node's centrality*. This represents a real-valued function tailored to provide a ranking [6] which identifies the most relevant nodes. For this purpose, some algorithms have been defined, such as Hypertext Induced Topic Search (HITS) algorithm [7] and Google's PageRank algorithm [8]. In our approach, we adopt Google's PageRank algorithm due to its efficiency [9]. In the following, we briefly introduce the PageRank algorithm and the random surfer model.

6.3.2.1 PageRank Algorithm

Definition 6.1 (PageRank) Let $I(v_i)$ be the set of vertices that link to a vertex v_i and let $deg_{out}(v_i)$ be the out-degree centrality of a vertex v_i. The PageRank of a vertex v_i, denoted by $PR(v_i)$, is provided by the following [8]:

$$PR(v_i) = d \left[\sum_{v_j \in I(v_i)} \frac{PR(v_j)}{deg_{out}(v_i)} \right] + (1-d)\frac{1}{|D|} \tag{6.1}$$

The constant d is called *damping factor*, assumed to be set to 0.85 [8]. The previous equation breaks down to one equation per node v_i with an equal number of unknown $PR(v_i)$ values. The PageRank algorithm tries to find out iteratively different PageRank values, which sum up to 1 ($sum_{i=1}^{n} PR(v_i) = 1$). The authors of the PageRank algorithm consider the use case of web surfing, where the user starts from a web page and randomly moves to another one through a web link. If the web surfer is on page v_j with a probability or a damping factor d, then the probability to change page v_i is $\frac{1}{deg_{out}(v_j)}$. The user could follow the links and teleport [8] to a random web page in V with $1-d$ probability. The described surfing model is a stochastic process, and W is a stochastic transition matrix, where node ranking values are computed as presented in the following:

$$\vec{PR} = d\left[W.\vec{PR}\right] + (1-d)\frac{1}{|D|}\vec{1} \tag{6.2}$$

The stochastic matrix W is defined as follows:

$w_{ij} = \frac{1}{deg_{out}(v_j)}$ if a vertex v_j is linked to v_i

$w_{ij} = 0$ otherwise

The notation $\vec{R}$ stands for a vector where its i_{th} element is $PR(v_i)$ (PageRank of v_i). The notation $\vec{1}$ stands for a vector having all elements equal to 1. The computation of PageRank values is done iteratively by defining a convergence stopping criterion ϵ. At each computation step t, a new vector $(\vec{PR}, t)$ is generated based on previous vector values $(\vec{PR}, t-1)$. The algorithm stops computing values when the condition $|(\vec{PR}, t) - (\vec{PR}, t-1)| < \epsilon$ is satisfied.

6.3.3 Security Correlation

6.3.3.1 Network Enrichment Using Passive DNS

A DNS sensor [10–13] is used to capture inter-server DNS communication in a passive DNS database. Afterward, the records of passive DNS stored in the database can be queried. We can benefit from a passive DNS database in many ways. For instance, we can know the history of a domain name, as well as the IP addresses that the domain is or was pointing to. We can also find out what domain names are hosted on a given name server or what domains are/have been pointing to a given IP address. There are a lot of use cases of passive DNS for security purposes (e.g., mapping criminal cyber-infrastructure [14], tracking spam campaigns, tracking malware command and control systems, detection of fast-flux networks, security monitoring of a given cyber-infrastructure, and botnet detection). In our context, we propose the correlation of ToGather intelligence with a passive DNS database to enrich, as shown in Fig. 6.5, the investigation of Android malware by: (1) finding suspicious domains that are pointing to a malicious IP address extracted from the analysis of a malware sample; (2) finding suspicious IP addresses that are resolved from a malicious domain that is obtained from the analysis of malware sample; (3) measuring the maliciousness magnitude of an IP. The maliciousness can be measured by counting the number of domains that resolve to this malicious IP address. Typically, these domains could be related to different malicious activities or a single one; (4) filtering outdated domain names: The passive DNS query generally returns timestamp information. ToGather could leverage timestamps to filter out old domain names that are no longer active.

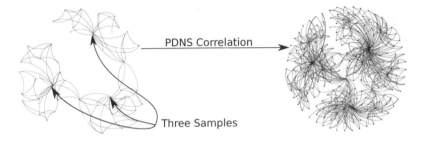

Fig. 6.5 Threat network with/without correlation

6.3.3.2   Threat Network Tagging

From Android malware samples, ToGather produces a threat network that summarizes their malicious activities. Afterward, ToGather detects and provides threat sub-networks if any. Besides, it helps prioritizing the actions to be taken to mitigate malicious activities using the PageRank algorithm. In this section, we go a step further toward the automatic investigation by leveraging other security feeds. Specifically, we aim at correlating threat networks with spam intelligence, reconnaissance intelligence, etc. The objective is to give a multi-dimensional view of the malicious activities that are related to the investigated Android malware family. Moreover, ToGather considers the correlation with network information from other platform malware; in the current setup, we correlate with PC malware from different operating systems.

PC Malware

ToGather tags every produced threat network by leveraging a database of network information extracted from PC malware VirusShare [15]. The malware database is continuously updated. The obtained data is identified by malware hash and its malware family. The latter helps identifying PC malware (and their families) that share network information with the Android threat network.

Spam

ToGather takes advantage of a spam database (30 Million records) to report the relationship between spamming campaigns and a given threat network. This information is precious for security analysts who are tracking spam campaigns.

Phishing

Similarly to spamming, we consider phishing activities in ToGather tagging. Phishing activities aim at stealing sensitive information using fake web pages that are similar to the known trusted ones. Typically, the attacker spreads phishing sites using malicious URLs. We extract only the domain name and store it in a phishing database (5 Million records).

Probing

ToGather considers tags of the threat network nodes if they are part of a probing activity. This presupposes the availability of a probing database (300 Million records) that contains IP addresses that have been part of scanning activities within the same epoch. Probing is derived from darknet traffic, and the probing IP addresses could be persisted in a probing database.

6.4   Experimental Results

In this section, we present the evaluation results of our proposed system. The goal of the evaluation is to assess the effectiveness of ToGather framework in terms of its ability to provide cyber-threat situational awareness from a set of Android malware samples.

6.4.1   Android Malware Dataset

In the evaluation, we use a real Android malware dataset from Drebin [16], a known dataset that contains samples labeled with their families. Drebin dataset [17] contains 5560 labeled malware samples from 179 families [17], as shown in Table 6.1. It is important to stress that Drebin contains all the samples of the MalGenome dataset [18]. As a ground truth for the malware labeling, we take the labels provided by Drebin since there are some differences between Genome and Drebin dataset labeling. For example, MalGenome recognizes different versions of DroidKungFu malware (1, 2, and 4), where Drebin has only DroidKungFu.

6.4.2   Implementation

We have implemented ToGather using *Python* programming language. In the static analysis, in order to perform reverse engineering of the *Dex* bytecode, we use *dexdump*, a tool provided with Android SDK. We extract the network information from the *Dex* disassembly using regular expressions. Beside, ToGather extracts network information from static text content in the APK file of Android malware.

Table 6.1 Dataset description by malware family

	Malware family	Number of samples
0	FakeInstaller	925
1	DroidKungFu	667
2	Plankton	625
3	Opfake	613
4	GinMaster	339
5	BaseBridge	330
6	Iconosys	152
7	Kmin	147

In the data enrichment phase, ToGather leverages the passive DNS database from Farsight Security Inc [11–13] using the company open source tools [19–21].

In dynamic analysis, a cornerstone of ToGather is the sandboxing system, which heavily influences the produced analysis reports. We use *DroidBox* [22], a well-established sandboxing environment based on the Android software emulator [23] provided by Google Android SDK [24] as presented in the previous chapter.

6.4.3   Drebin Threat Network

In this section, we present the results of applying ToGather framework on the samples of Drebin dataset with all the 179 families. Figure 6.6 depicts the threat network information (domain names and IP addresses) of Drebin dataset, where a

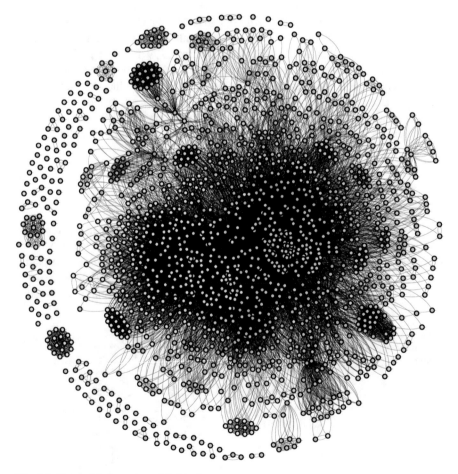

Fig. 6.6 Network information of drebin dataset

different color represents each family. Although the threat networks are not very clear visually, we could distinguish some connected communities with the same nodes' colors, i.e., the same malware family. This initial observation enhances the need for a community detection module in ToGather. The community here is a set of graph nodes that are highly connected even though they share some links with external nodes. In Fig. 6.7, we consider only domain names; here, we can distinguish more sub-threat networks having nodes from the same malware family. We choose to filter all IP addresses for Drebin dataset due to observations during the experimentation process: (1) Some malware samples contain a significant number of IP addresses; exceeding, in some cases, 100 IPs such as Plankton sample with MD5 hash 3f69836f64f956a5c00aa97bf1e04bf2 . The adversary could deceive the investigator by overwhelming the app with unused IP addresses along with used ones. (2) A big portion of the IP addresses are part of cloud infrastructure; we filter

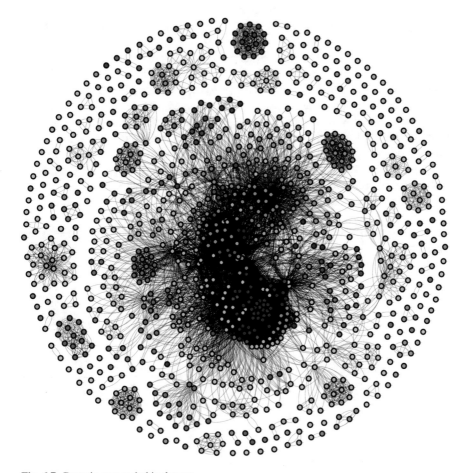

Fig. 6.7 Domain names drebin dataset

most of the public ones, but there are plenty of less known infrastructures in remote countries. (3) In most cases, the adversary utilizes domains for malicious activity due to the low cost and flexibility compared to IP addresses. In this experimentation, we consider only domain names, but the security analyst could include IP addresses when needed.

Using all Drebin dataset (179 malware families) to produce the threat network is an extreme use case for ToGather framework; using only few malware families represents a typical use case when we aim to investigate the threat network relations. However, even with the whole Drebin dataset, ToGather, as presented in Fig. 6.7, shows promising results, where we notice sub-threat networks with/without links to other nodes. By considering only domain names in Fig. 6.7, it is noticeable that the size of the threat network significantly decreases by removing IP addresses; typically, there are substantially more domains than IP addresses in the Android apps . However, this is due to the extensive whitelisting of domains compared to IPs (more than 1 million domain) and the size of the Drebin dataset. At this stage, we do not present the community detection and page ranking on the threat network; this will be conducted on a one-family use case in Sect. 6.4.4. ToGather leverages different malicious datasets, as previously described in Sect. 6.3.3.2, to tag the nodes of the produced threat network. Table 6.2 depicts the diverse malicious activities of the nodes from Drebin threat network. First, the table shows the top PC malware families, which share network information with the Drebin threat network. For family names, we adopt the *Kaspersky* malware family naming as our ground truth. Besides, Fig. 6.8 shows the percentage of each malicious type in the Drebin threat network. The result indicates that 56% of the shared nodes have a spamming activity, 40% are related to PC malware, 3% scanning, and 1% phishing activities. Notice that the previous percentages are only from the shared nodes and not from all the threat networks.

Table 6.2 Drebin dataset tagging results

#	Family	Hits
1	Agent[a]	1268
2	VBNA	283
3	Adload	152
4	EgroupDial	121
5	TrustAsia	120
6	Vobfus	88
7	KuPlays	74
8	Pipibo	72
9	Sality	62

[a] Kaspersky naming

Fig. 6.8 Drebin dataset
tagging distribution results

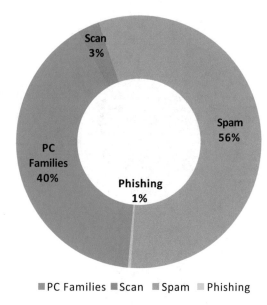

6.4.4 **Family Threat Networks**

In this section, we present the results of ToGather in its typical usage scenario
where malware samples from the same family are analyzed. Figure 6.9 shows the
steps of generating threat networks from the DroidKungFu family sample. First,
ToGather produces the threat network, including network information collected
from the DroidKungFu analysis and Passive DNS correlation, as shown in
Fig. 6.9a. Afterward, ToGather filters the whitelist network information. The
results, shown in Fig. 6.9b, depict bright separated sub-threat networks without
applying the community detection algorithm. This could be an insightful result for
the security practitioner, especially that this sub-threat network contains network
information exclusively from some samples. ToGather goes a step ahead by
applying both community detection (resolution hyper-parameter $r = 3$) and
page ranking algorithms (damping factor $d = 0.85$ and stopping criterion $\epsilon =
0.001$) as hyper-parameters to divide the network and rank the importance of the
nodes, respectively. The result consists of multiple sub-threat networks, with high
interconnection and low intra-connection, representing the cyber-infrastructures of
DroidKungFu malware family.

Figure 6.10 shows the results provided by ToGather when using Android malware samples from BaseBridge family. Similarly, after the filtering operation, we
could easily distinguish small sub-threat networks. In some cases, the community
detection task could be optional due to the clear separation between the sub-threat
networks. For instance, Fig. 6.11 depicts the threat networks for GinMaster,
Adrd, and Plankton Android malware families before and after the community
detection task. Here, Adrd family has multiple sub-threat networks without the

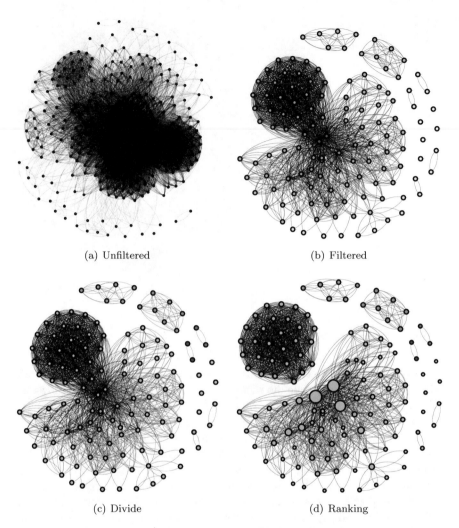

(a) Unfiltered   (b) Filtered

(c) Divide   (d) Ranking

Fig. 6.9 DroidKungFu malware threat network. (**a**) Unfiltered. (**b**) Filtered. (**c**) Divide. (**d**) Ranking

need for the community detection function since it does not affect the results much. In the case of Plankton, it is necessary to detect and extract the sub-threat network.

Tables 6.3 and 6.4 show the top PC malware families and samples that share the network information with BaseBridge and DroidKungFu threat networks. An essential factor in the correlation is the explainability, where we could determine which network information is shared between the Android malware and PC malware. This could help the security investigator to track the other dimension of the adversary cyber-infrastructure.

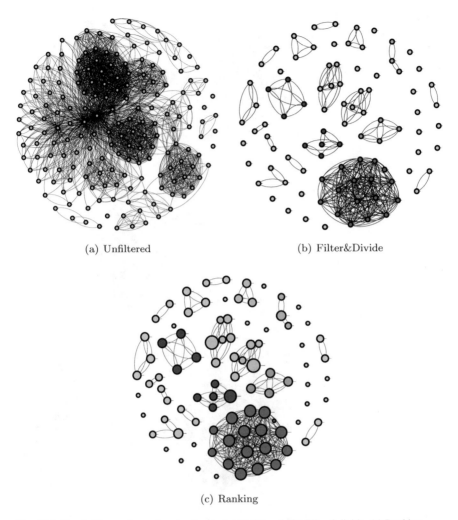

(a) Unfiltered   (b) Filter&Divide

(c) Ranking

Fig. 6.10 Basebridge malware threat network. (**a**) Unfiltered. (**b**) Filter&divide. (**c**) Ranking

In addition to the PC malware tagging, we correlate with other cyber malicious activity datasets over the Internet. Figure 6.12 presents the malicious activities of DroidKungFu and BaseBridge families that are related to their threat network. Here, we find that both families could be part of a spam campaign while having some scanning activity. Notice that these results represent a fraction of the actual activity because of limited datasets.

Fig. 6.11 Android families from drebin dataset. (**a**) Ginmaster (1). (**b**) Adrd (1). (**c**) Plankton (1). (**d**) Ginmaster (2). (**e**) Adrd (2). (**f**) Plankton (2)

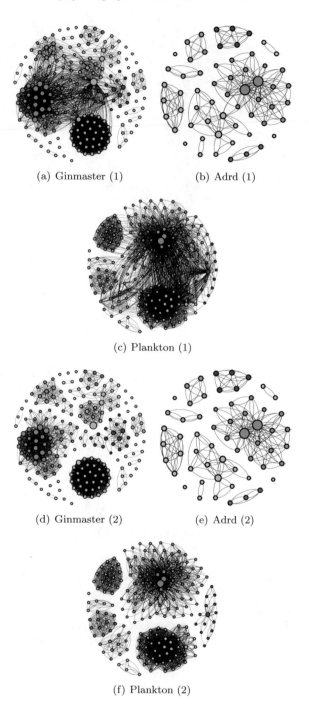

(a) Ginmaster (1)   (b) Adrd (1)

(c) Plankton (1)

(d) Ginmaster (2)   (e) Adrd (2)

(f) Plankton (2)

Table 6.3 Top PC malware related to basebridge family

#	Sample	Hits		#	Family	Hits
1	ed7621ec4d[a]	2		1	Agent[b]	23
2	e3bc76d14c	2		2	Vobfus	21
3	503902c503	1		3	EgroupDial	13
4	bd9b87869b	1		4	Badur	9
5	8e0cf0a1ba6	1		5	LMN	7
6	f8a5cac12dc	1		6	WBNA	4
7	14db95e5f6	1		7	Pipibo	2
8	9b5b576ef3	1		8	Blocker	2
9	2ec2abc28d	1		9	Virut	2

[a] MD5 hash first 10 chars   [b] Kaspersky naming

Table 6.4 Top PC malware related to DroidKungFu family

#	Sample	Hits		#	Family	Hits
1	74529155cc[a]	3		1	Agent[b]	33
2	bd5a9f768cf	2		2	Adload	24
3	259a244ab1	2		3	TrustAsia	13
4	52da75225	1		4	KuPlays	11
5	11786afada	1		5	Pipibo	8
6	ad5e6d577b	1		6	FangPlay	5
7	9f4215bfc3	1		7	StartPage	4
8	3c76ff67d0	1		8	Injector	4
9	117f21550	1		9	Turbobit	4

[a] MD5 hash first 10 chars   [b] Kaspersky naming

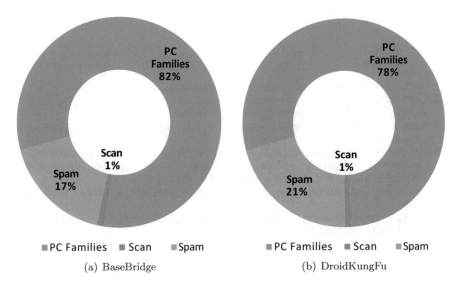

(a) BaseBridge   (b) DroidKungFu

Fig. 6.12 Maliciousness tagging per family. (**a**) BaseBridge. (**b**) DroidKungFu

6.5 Summary

In this chapter, we presented the ToGather framework, a set of techniques, tools, and mechanisms as well as security feeds bundled together in order to achieve situational awareness about Android malware threats automatically. ToGather leverages state-of-the-art graph partitioning algorithms and multiple security feeds to produce insightful, granular, as well as actionable intelligence about malicious cyber-infrastructures related to Android malware samples. We evaluated ToGather on real malware from the Drebin Dataset [16]. The results show promising insights about cyber-infrastructures of Android malware families. The produced threat networks could show one side of the adversary infrastructure, which is the Android malware one; this side could lead to a larger threat network. Furthermore, all the results can be extracted automatically and periodically from a feed of Android malware samples belonging to one or various families. This requires fixing the hyper-parameters related to the used algorithms of the community detection, and page ranking, as we did in our experimentation.

In this chapter as well as in Chaps. 4 and 5, we propose Android malware fingerprinting systems that target the workstation category in the taxonomy proposed in Chap. 2. In the next chapter, we propose a portable Android malware detection system that targets all the deployment categories (mentioned in Chap. 2). More specifically, this involves a detection system that is versatile enough to be efficiently deployed on high-end servers as well as on IoT devices such as Raspberry PI boards.

References

1. Alexa Top Sites - http://www.alexa.com/topsites. Accessed Nov 2016
2. Quantcast Sites - https://tinyurl.com/gmd577y. Accessed Nov 2016
3. Tracemyip - http://tools.tracemyip.org/. Accessed Jan 2016
4. Amazon IP Space - https://docs.aws.amazon.com/general/latest/gr/aws-ip-ranges.html. Accessed Nov 2016
5. V.D. Blondel, J.-L. Guillaume, R. Lambiotte, E. Lefebvre, Fast unfolding of communities in large networks. *Journal of statistical mechanics: theory and experiment*, (10), P10008 (2008)
6. S.P. Borgatti, Centrality and network flow. Soc. Netw. **27**(1), 55–71 (2005)
7. J.M. Kleinberg, Authoritative sources in a hyperlinked environment. J. ACM **46**(5), 604–632 (1999)
8. S. Brin, L. Page, The anatomy of a large-scale hypertextual web search engine. Comput. Netw. **30**(1–7), 107–117 (1998)
9. R.W. Nidhi Grover, Comparative analysis of PageRank and HITS algorithms. *International Journal of Engineering Research & Technology (IJERT)* **1**(8), 1–15 (2012).
10. F. Weimer, Passive DNS replication (2005). http://www.enyo.de/fw/software/dnslogger/first2005-paper.pdf
11. Security information exchange (SIE), farsight security inc. - https://www.farsightsecurity.com. Accessed March 2020
12. Farsight Security Inc. DNSDB. - https://www.dnsdb.info. Accessed March 2020

13. Farsight Security Inc. DNSTable. - https://github.com/farsightsec/dnstable/blob/master/man/dnstable-encoding.5.txt. Accessed March 2020
14. M. Antonakakis, R. Perdisci, D. Dagon, W. Lee, N. Feamster, Building a dynamic reputation system for DNS, in *19th USENIX Security Symposium, Washington, DC, USA, August 11–13, 2010, Proceedings* (2010), pp. 273–290
15. VirusShare malware repository- https://virusshare.com/. Accessed Aug 2018
16. D. Arp, M. Spreitzenbarth, M. Hubner, H. Gascon, K. Rieck, DREBIN: effective and explainable detection of android malware in your pocket, in *21st Annual Network and Distributed System Security Symposium, NDSS 2014, San Diego, California, USA, February 23–26, 2014* (2014)
17. Drebin Android Malware Dataset - https://user.informatik.uni-goettingen.de/~darp/drebin/. Accessed Jan. 2015
18. Android Malware Genome Project - http://www.malgenomeproject.org/. Accessed Jan. 2015
19. Farsight Security Inc. Immutable sorted string table library (mtbl) - https://github.com/farsightsec/mtbl. Accessed Mar 2020
20. Farsight Security Inc. Nmsg library. - https://github.com/farsightsec/nmsg. Accessed Mar 2020
21. Farsight Security Inc. Pynmsg library. - https://github.com/farsightsec/nmsg. Accessed Mar 2020
22. DroidBox - https://github.com/pjlantz/droidboxh. Accessed May 2016
23. Android Emulator - https://developer.android.com/studio/run/emulator. Accessed Feb 2016
24. Android Software Developer Kit (SDK) - https://developer.android.com/studio/. Accessed June 2019

Chapter 7
Portable Supervised Malware Fingerprinting Using Deep Learning

In this chapter, we propose MalDozer, an innovative and efficient framework for Android malware detection, leveraging sequence mining via neural networks. MalDozer focuses on portable malware detection based on applying supervised machine learning on static analysis features in contrast to Cypider, presented in Chap. 4, in which we propose an unsupervised system based on static analysis features. While Cypider provides a framework for malware clustering, aimed at market level app analysis. MalDozer provides an efficient malware detection to allow the deployment inside resource-constrained devices. MalDozer framework is based on an artificial neural network that takes, as input, the raw sequences of API method calls, as they appear in the DEX file, to enable malware detection and family attribution. MalDozer can automatically recognize malicious patterns using only the sequences of raw method calls in the assembly code.

More precisely, the input of MalDozer is the sequences of the API method calls as they appear in the *DEX* file, where a sequence represents Android app. First, we map each method in the sequence invocation to a fixed-length high-dimensional vector that semantically represents the method invocation [1] and replaces the sequence of the Android app methods by a sequence of vectors. Afterward, we feed the sequence of vectors to a neural network with multiple layers.

7.1 Threat Model

We position MalDozer as an anti-malware system that detects Android malware and attributes it to a known family with high accuracy and minimal false positive and false negative rates. We assume that the analyzed Android apps, whether malicious or benign, are developed mainly in Java or any other language that is translated to DEX bytecode. Therefore, Android apps developed by other means, e.g., web-based, are out of the scope of the current design of MalDozer. Also, we assume that apps' core functionalities are in DEX bytecode and not in C/C++ native code [2], i.e., the attacker is mainly using the DEX bytecode for the malicious payload. Furthermore, we assume that MalDozer detection results are not affected by malicious activities. In the case of a server, Android malicious apps are assumed to not modify the server system. However, in the case of deployment on infected mobiles or IoT devices, MalDozer should be protected from malicious activities to avoid tampering its results.

7.2   Usage Scenarios

The effectiveness of MalDozer, i.e., its high accuracy, makes it a suitable choice for malware detection in market level deployment (Taxonomy in Chap. 2), especially that its update only requires very minimal manual intervention. We only need to train MalDozer model on new samples without a *feature engineering*, since MalDozer can automatically extract and learn relevant malicious and benign features during the training. Notice that MalDozer could detect unknown malware based on our evaluation as presented in Sect. 7.4. Furthermore, due to the efficiency of MalDozer, it could be deployed on mobile devices such as phones and tablets. As for mobile devices, MalDozer acts as the detection component in the anti-malware app inside Android phones, where the goal is to check the maliciousness likelihood of new apps. Family attribution is very handy when detecting new malware apps. Indeed, MalDozer helps the anti-malware system to take the necessary precautions and actions based on the malware family, which could involve specific malicious threats such as ransomware.

7.3   Methodology

In this section, we present MalDozer framework and its components (Fig. 7.1). MalDozer has a simple design, where a minimalistic preprocessing is employed to obtain the assembly code methods. As for the feature extraction (representation learning) and detection/attribution, they are based on the actual neural network. This permits MalDozer to be very efficient with fast preprocessing and neural network execution. Since MalDozer is based on supervised machine learning, we first need to train our model. Afterward, we deploy the trained model along with a preprocessing procedure on the targeted devices. Notice that the preprocessing procedure is common between the training and the deployment phases to ensure the correctness of the detection results (Fig. 7.1).

1. Extraction of API Method Calls
MalDozer workflow extracts the sequences of API calls from Android app packages, in which we consider only *DEX* file. We disassemble the classes.dex to

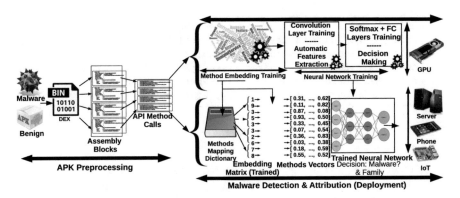

Fig. 7.1 Approach overview

```
android/net/ConnectivityManager
android/net/ConnectivityManager
android/telephony/SmsManager
android/telephony/SmsManager
android/location/LocationManager
android/location/LocationManager
```

Fig. 7.2 Android API from a Malware sample

produce Dalvik VM assembly. Our goal is to model the Dalvik assembly to keep the maximum raw information with minimum noise. Notice here that we could use Android APIs (such as android/net/ConnectivityManager in Fig. 7.2) instead of permission to have a granular view that helps distinguishing malware apps.

However, quantifying Android API could be challenging because there are plenty of common API calls shared between apps. Some solutions tend to filter only sensitive APIs and use them for detection. In this case, we require a manual categorization of sensitive APIs. Moreover, Android API gives an abstract view of the actual malicious activity that could hinder malware detection. For this reason, we leverage Android API method calls as android/net/ConnectivityManager;-> getNetworkInfo in Fig. 7.3. By doing so, the proposed malware detector will have a more granular view of the app activity. In our case, we address this problem from another angle; we treat Android apps as a sequence of API method calls. We consider all the API calls with no filtering, where their calling order is part of the information we use to identify malware. It represents the temporal relationship between two API method calls (in a given basic block) and defines the intended sub-tasks of the app. The sequence of API method calls preserves the temporal relationship over the individual basic blocks of the linear disassembly and ignores the order of the basic blocks. The obtained result is a merged sequence (Fig. 7.1).

```
android/net/ConnectivityManager;->getNetworkInfo
android/net/ConnectivityManager;->getAllNetworkInfo
android/telephony/SmsManager;->sendTextMessage
android/telephony/SmsManager;->sendMultipartTextMessage
android/location/LocationManager;->getLastKnownLocation
android/location/LocationManager;->getBestProvider
```

Fig. 7.3 Granular view using API method calls

In other words, a *DEX* file, denoted by cd, is composed of a set of n compiled Java classes, $cd = \{cl_1, \ldots, cl_n\}$. Each Java class cl_i is, in turn, composed of a set of m methods, which are basic blocks, $cl_i = \{mt_1^i, \ldots, mt_m^i\}$. By going down to the API method level, mt_j^i is a sequence of k API method calls. Formally $mt_j^i = (P_1^{i,j}, \ldots, P_k^{i,j})$, where $P_l^{i,j}$ is the lth API method call in method mt_j^i.

2. Dictionary Mapping of API Method Calls

In this step, we map the sequences of API method calls that are in an Android app to the corresponding identifiers. More precisely, we replace each API method with an identifier, resulting in a sequence of numbers. We also build a dictionary that maps each API call to its identifier. Notice that in the current implementation, the mapping dictionary is deployed with the learning model to map the API calls of the analyzed apps. In the deployment, we might find unknown API calls related to third party libraries. To overcome this problem: (1) We consider a large training dataset that covers most of the API calls. (2) In the deployment phase, we replace unknown API calls with fixed identifiers. Afterward, we unify the length of the sequences L (hyper-parameter) and pad a given sequence with zeros if its length $l < L$.

3. Unification of the Sequences' Size

The length of the sequences varies from one app to another. Hence, it is important to unify the length of the sequences. Two cases are depending on the length of the sequence and the hyper-parameter. We choose a uniform sequence size as follows: (1) If the length of a given sequence is greater than the uniform sequence size L, we take only the first L items to represent the apps. (2) In case the length of the sequence is less than L, we pad the sequence with zeros. It is important to mention that the uniform sequence size hyper-parameter influences the accuracy of MalDozer. A simple rule is that the larger is the size, the better it is, but this will require a lot of computation power and a long time to train the neural network.

4. Generation of the Semantic Vectors

The identifier in the sequences needs to be shaped to fit as input to our neural network. The issue could be solved by representing each identifier by a vector. The question that arises is *how are such vectors produced?* A straightforward solution is to use one-hot vectors, where a vector has one in the interface value row, and zero in the rest. Such a vector is very sparse because its size is equal to the number of API calls, which makes it impractical and computationally prohibitive for the training and the deployment. To address this issue, we resort to dense vectors. These vectors are semantically related, and we could express their relation by computing a distance. The smaller the distance is, the more related the vectors are (i.e., API calls). We describe word embedding in Sect. 7.3.1. The output of this step is sequences of vectors for each app that keeps the order of the original API calls; each vector has a fixed size K (hyper-parameter).

5. Prediction Using a Neural Network
The final component in MalDozer framework is the neural network, which is composed of several layers. The number of layers and the complexity of the model are hyper-parameters. However, we aim to keep the neural network model as simple as possible to reduce the execution time during its deployment, especially on IoT devices. In our design, we rely on the convolutional layers [3] to automatically discover the pattern in the raw method calls. The input to the neural network is a sequence of vectors, i.e., a matrix of $L \times K$ shape. In the training phase, we train the neural network parameters (layers weight) based on the app vector sequence and its labels: (1) malware or benign for the detection task, and (2) malware families for the attribution task. In the deployment phase, we extract the sequence of methods and use the embedding model to produce the vector sequence. Finally, the neural network takes the vector sequence to decide about the given Android app.

7.3.1   MalDozer Method Embedding

The neural network takes vectors as input. Therefore, we represent our Android API method calls as vectors. As a result, we formalize an Android app as a sequence of vectors with a fixed size (L). We could use one-hot vectors. However, their size is the number of unique API method calls in our dataset. This makes such a solution not scalable to a large-scale training. Also, the word embedding technique outperforms the results of the one-hot vector technique in our case [1, 3, 4]. Therefore, we seek a compact vector, which also has semantic value. To fulfill these requirements, we choose the word embedding techniques, namely word2vec [1] and GloVe [4]. Our primary goal is to have for each Android API method a dense vector; the vector' values are learned from the method contexts in a large dataset of Android apps. Thus, in contrast to one-hot vectors, each word embedding vector contains a numerical summary of the Android API call semantic representation.

Moreover, the learned API call vectors have semantic relationships to each other in terms of functionality, i.e., developers tend to use specific API method calls in the same context. In our case, we learn these vectors from our dataset that contains benign and malicious apps by using word2vec [1]. The latter is a computationally efficient predictive model based on learning word embedding vectors, which are applied in our case to raw Android API method calls. The output obtained from training the embedding word model is a matrix $K \times A$, where K is the size of the embedding vector, and A is the number of unique Android API method calls.

Both K and A are hyper-parameters; we use $K = 64$ in all our models. We choose $K = 64$ because: (1) it is a common practice in NLP literature to use $K \in \{32, 64, 128, \ldots\}$ values for word embedding vectors sizes. (2) The value 64 is a good tradeoff between the efficiency, which is required in mobile deployment, and the effectiveness compared to the use of 32 and 128 embedding vector sizes. In the deployment phase (Fig. 7.1), MalDozer uses the word embedding model and looks up for each API method call identifier to find the corresponding embedding vector.

7.3.2   MalDozer Neural Network

MalDozer neural network is inspired by Kim [3], where the authors use a neural network for sentence classification tasks such as sentiment analysis. The proposed architecture shows high results and outperforms many of state-of-the-art benchmarks with a relatively simple neural network design. We raise the questions: *Why could such a Natural Language Processing (NLP) model be useful in Android malware detection?* and *why do we choose to build it on top of this design [3]?* We formulate our answers as follows: (1) NLP is a challenging field where we deal with text. So, there is an enormous number of vocabularies; also, we could express the same meaning in different ways. Besides, the same semantics could be expressed with different combinations of words, which is the equivalent of code obfuscation in natural language processing.

In our context, we deal with sequences of Android API method calls and want to find the combination of method calls patterns, which produces the same (malicious) activity. We use API method calls as they appear in the bytecode. Indeed, there is a temporal relationship between API methods in basic blocks. Nevertheless, the extraction process neglects the order among blocks and only considers the order inside the code blocks. By analogy to NLP, blocks are sentences, and the API method calls are words. Further, an app (paragraph) is a list of basic blocks (unordered sentences). Malware detection using API method calls looks easier compared to the NLP one because of the huge difference in the vocabulary, i.e., the number of Android API method calls is significantly less than the number of words in natural language. Also, combinations in natural language are much more complex compared to Android API calls. (2) We choose to use this model due to its efficiency and ability to run on resource-constrained devices. Table 7.1 depicts the neural network architecture of MalDozer's detection and attribution tasks. Since both networks are very similar, the only notable difference is in the output layer.

In the detection task, we need only one neuron in the output layer because the network decides whether the app is malware or not. As for the attribution task, there are multiple neurons, one for each Android malware family. Having the same architecture for the detection and attribution makes the development and the evaluation of a given design simpler. Because the network architecture achieves good results in one task, it will have very similar results in the other one. As

Table 7.1 MalDozer malware neural network

#	Layers	Options	Active
1	Convolution	Filter=512, FilterSize=3	ReLU
2	MaxPooling	–	–
3	FC	#Neurons=256, Dropout=0.5	ReLU
4	FC	#Neurons={1,#Families[a] }	Softmax

[a]The number of malware families in the training dataset

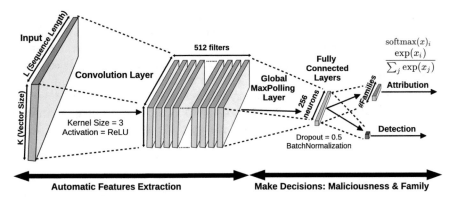

Fig. 7.4 Neural network architecture

presented in Fig. 7.4, the first layer is a convolution layer [3] with Rectified Linear Unit (ReLU) activation function ($f(x) = max(0, x)$). Afterward, we use the global max pool [3] and connect it to a fully connected layer. Notice that in addition to Dropout [5] used to prevent over-fitting, we also utilize Batch Normalization [6] to improve our results. Finally, we have an output layer, where the number of neurons depends on the detection or attribution tasks.

7.3.3 Implementation

In this section, we present the software and hardware components of MalDozer implementation.

Software

We implement MalDozer using *Python* and *Bash* scripting languages. First, Python zip library extracts the *DEX* file from the *APK* file. We use *dexdump* command-line to produce the assembly from DEX file. *Dexdump* is available through the Android SDK, but in the case of Raspberry PI, we build it from its source code. Regular expressions are employed to extract API method calls from the assembly. To develop the neural network, we use Tensorflow library [7]. Notice that there is no optimization in the preprocessing; in the runtime evaluation, we use only a single thread app.

Table 7.2 Hardware
specifications

	Server (1/2)	Laptop	RPI2
GPU	TITAN X /no	No	No
CPU	Intel E5-2630	Intel T6400	ARM Core A7
RAM	128GB	3GB	1GB

Hardware

To evaluate the efficiency of MalDozer, we evaluate multiple types of hardware,
as shown in Table 7.2, starting from servers to *Raspberry PI* [8]. For training,
the Graphics Processing Unit (GPU) is a vital component because the neural
network training needs immense computational power. The training takes hours
under *NVIDIA TitanX*. However, the deployment could be virtually on any device,
including IoT devices (such as Raspberry Pi). To this end, we consider *Raspberry PI*
as an IoT device because it is one of the hardware platforms supported by Android
Things [9]. We also use low-end laptops in our evaluation, as shown in Table 7.2.

7.4 Evaluation

In this section, we conduct our evaluation using different datasets that primarily
cover the following performance aspects: (1) *Detection Performance*: We evaluate
how effectively MalDozer can distinguish between malicious and benign apps
in terms of F1-measure, precision, recall, and false-positive rate. (2) *Attribution
Performance*: We evaluate how effectively MalDozer can correctly attribute a given
malicious app to its malware family. (3) *Runtime Performance*: We measure the
preprocessing and the detection runtime on different types of hardware.

7.4.1 Datasets

In our evaluation, we have two main tasks: (1) Detection, which aims at checking
if a given app is malware or not, (2) Attribution, which aims at determining the
family of the detected malware. We conduct the evaluation experiments under two
types of datasets: (1) Mixed dataset, which contains malicious apps and benign
apps, as presented in Table 7.3. (2) Malware dataset, which contains only malware,
as shown in Table 7.4. As for the malware dataset, we leverage reference datasets
such as *Malgenome* [10] and *Drebin* [11]. We also collect two other datasets from
different sources, e.g., *virusshare.com, Contagio Minidump* [12]. The total number
of malware samples is $33K$, including Malgenome and Drebin datasets. As for the
attribution task, we use only malware from the previous datasets, where each family
has at least 40 samples, as presented in Tables 7.12, 7.13, and 7.14. To this end, we
propose MalDozer dataset, as in Table 7.12, which contains 20K malware samples

Table 7.3 Datasets for detection task

Dataset	#Malware	#Benign	Total
Malgenome	1258	37,627	38,885
Drebin	5555	37,627	43,182
MalDozer	20,089	37,627	57,716
All	33,066	37,627	70,693

Table 7.4 Datasets for attribution task

Dataset	#Malware	#Family
Malgenome	985	9
Drebin	4661	20
MalDozer	20,089	32

Table 7.5 Detection on Malgenome dataset

	F1%	P%	R%	FPR%
2-Fold	99.6600	99.6620	99.6656	0.06
3-Fold	98.1926	98.6673	97.9812	1.97
5-Fold	99.8044	99.8042	99.8045	0.09
10-Fold	**99.8482**	**99.8474**	**99.8482**	**0.04**

The bold values represent the best values

from 32 malware families. We envision to make MalDozer dataset available upon request for the research community. The benign app samples have been collected from *Playdrone* dataset [13]. We leverage the top $38K$ apps that are ranked by the number of downloads.

7.4.2 Malware Detection Performance

We evaluate MalDozer on different cross-validation settings, two, three, five, and ten-fold, to examine the detection performance under different training/test set percentages $(50\%, 66\%, 80\%, 90\%)$ from the actual dataset (10 training epochs). Table 7.5 depicts the detection results on Malgenome dataset. MalDozer achieves excellent results, F1-Score $= 99.84\%$, with a small *False Positive Rate* (FPR), 0.04%, despite the unbalanced dataset, where benign app samples are the most dominant in the dataset. The detection results are similar under all cross-validation settings. Table 7.6 presents the detection results on Drebin dataset, which are very similar to the Malgenome ones. MalDozer reaches F1-Score $= 99.21\%$, with FPR $= 0.45\%$. Similar detection results are shown in Table 7.7 on MalDozer dataset (F1-Score $= 98.18\%$ and FPR $= 1.15\%$). Table 7.8 shows the results related to all datasets, where MalDozer achieves a good result (F1-Score $= 96.33\%$). However, it has a higher false positive rate compared to the previous results (FPR $= 3.19\%$). This leads us to manually investigate the false positives. We discover, by correlating with *virusTotal.com*, that several false positive apps are already detected by many vendors as malware.

Table 7.6 Detection on Drebin dataset

	F1%	P%	R%	FPR%
2-Fold	98.8834	98.9015	98.9000	0.13
3-Fold	99.0142	99.0130	99.01579	0.51
5-Fold	99.1174	99.1173	99.1223	0.31
10-Fold	**99.2173**	**99.2173**	**99.2172**	**0.45**

The bold values represent the best values

Table 7.7 Detection on MalDozer dataset

	F1%	P%	R%	FPR%
2-Fold	96.8576	96.9079	96.8778	1.01
3-Fold	97.6229	97.6260	97.6211	2.00
5-Fold	97.7804	97.7964	97.7753	2.25
10-Fold	**98.1875**	**98.1876**	**98.1894**	**1.15**

The bold values represent the best values

Table 7.8 Detection on all dataset

	F1%	P%	R%	FPR%
2-Fold	96.0708	96.0962	96.0745	2.53
3-Fold	95.0252	95.0252	95.0278	4.01
5-Fold	96.3326	96.3434	96.3348	2.67
10-Fold	**96.2958**	**96.2969**	**96.2966**	**3.19**

The bold values represent the best values

7.4.2.1   Unknown Malware Detection

Although MalDozer demonstrates very good detection results, some questions still arise: (1) *Can MalDozer detect samples of unknown malware families?* and (2) *How many samples are needed for a given family to achieve a good accuracy?* To answer these questions, we conduct the following experiment on Drebin mixed dataset (Malware + Benign), where we focus on top malware families, i.e., BaseBridge, DroidKungFu, FakeInstaller, GinMaster, Opfake, and Plankton. For each family, we train (5 epochs) our model on a subset dataset, which does not include samples of that family. These samples are used as a test set. Afterward, we train with few samples from the family and evaluate the model on the rest of the sample in that family. Progressively, we add more samples to the training and assess the accuracy of our model on detecting the rest of the family samples. Answering the above questions: *(1) Can MalDozer detect unknown malware family samples?* Yes, Fig. 7.5 shows the accuracy versus the number of samples in the training dataset. We see that MalDozer (zero samples versus accuracy) could detect the unknown malware family samples without previous training. The accuracy varies from 60% to 90%. *(2) How many samples for a given family to achieve a good accuracy?* MalDozer needs only about 10 to 20 samples to reach 90% (Fig. 7.5). In the case of DroidKungFu, MalDozer needs 20 samples to reach 90%. Considering only 10 to 20 samples from a malware family is a rather small number to obtain quality results. This varies from a malware family to another due to: (1) the similarity of

Fig. 7.5 Evaluation of
unknown malware detection.
(**a**) BaseBridge. (**b**)
DrKungFu. (**c**) FakeInst. (**d**)
GinMaster. (**e**) Opfake. (**f**)
Plankton

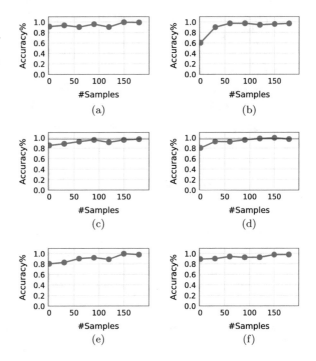

the malicious pattern of the new family compared known patterns. The higher the similarity to existing malware, the better the detection performance with (sometimes without any family sample in the training dataset) minimum samples in the training dataset. (2) Some new malware families tend to have simple patterns in their payload. Therefore, the learning system needs only a few samples to grasp the patterns of the whole family.

7.4.2.2 Resiliency Against API Evolution Over Time

As we have seen in the previous section, MalDozer could detect new malware samples from unknown families using samples from *Drebin* dataset collected in the period of 2011/2012. We aim to answer another important question: *Can MalDozer detect malicious and benign apps collected in different years?* To answer this question, we evaluate MalDozer on four datasets collected from [14] spanning across four consecutive years: 2013, 2014, 2015, and 2016. We take five malicious apps and five benign apps from the samples of each year. Then, we train MalDozer in 1 year dataset and test it on the rest of the datasets. The obtained results show that MalDozer detection is more resilient to API evolution over time compare to the result reported in [15], as presented in Fig. 7.6. Starting with 2013 dataset (Fig. 7.6a), we train MalDozer on 2013 samples and evaluate it on 2014, 2015, and 2016 ones. We notice a high detection rate in the 2014 dataset since it is collected

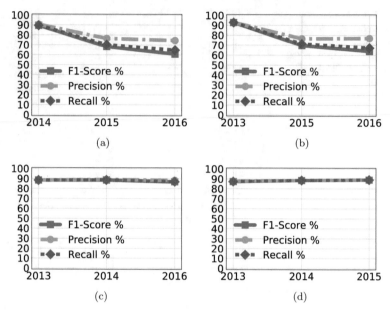

Fig. 7.6 Detection versus time. (**a**) 2013 dataset. (**b**) 2014 dataset. (**c**) 2015 dataset. (**d**) 2016 dataset

in the consecutive year of the training dataset. However, the detection rate decreases in 2015 and 2016 datasets, but it is above an acceptable detection rate (F1-Score = 70%). Similarly, we obtained the results of the 2014 dataset, as depicted in Fig. 7.6b. Also, training MalDozer on 2015 or 2016 datasets exhibits excellent results under all the datasets collected in other years, where we reach F1-Score from 90 to 92.5%.

7.4.2.3   Resiliency Against Changing the Order of API Methods

In the following, we evaluate the robustness of MalDozer against changes in the order of API method calls. Such changes may occur for various reasons, such as: (1) We could use different disassembly tools in the production, (2) A malware developer could repackage the same malicious app multiple times. The previous scenarios could lead to losing the temporal relations among the API calls. In the case of malware developer, she/he will be limited by keeping the same malicious semantics in the app. To validate the robustness of MalDozer against such methods that alter the order, we conduct the following experiment: First, we train our model on the training dataset. Afterward, we randomly shuffle the sequence of API method calls in the test dataset. We divide the testing app sequence into N blocks, then shuffle them and evaluate the F1-Score. We repeat until N is equal to the number of sequences, i.e., one API call in each block. The result of this experiment is shown in Fig. 7.7. The latter depicts the F1-Score versus the number of blocks, starting

Fig. 7.7 Shuffle rate versus
F1-score

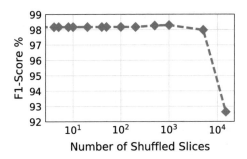

Table 7.9 Attribution on
Malgenome

	F1%	P%	R%
2-Fold	98.9834	99.0009	98.9847
3-Fold	98.9910	99.0026	98.9847
5-Fold	99.0907	99.1032	99.0862
10-Fold	**99.1873**	**99.1873**	**99.1878**

The bold values represent the best values

with four blocks and ending with 15K blocks, where each block contains one API
call. Figure 7.7 demonstrates the resiliency of MalDozer against changing the order
of API method calls. We observe that even with completely random individual API
method calls, MalDozer still achieves 93% F1-score.

7.4.3 Family Attribution Performance

Family attribution is an important task for Android security, where MalDozer
distinguishes itself from the existing malware detection solutions, since only few
solutions provide this functionality. Starting with Malgenome dataset (Table 7.9),
MalDozer achieves a very good result, i.e., F1-Score of 99.18%. Similarly, Mal-
Dozer reaches an F1-Score of 98% on Drebin dataset (Table 7.10). The results per
malware family attribution performance for Malgenome and Drebin are presented
in Tables 7.13 and 7.14. MalDozer achieves good results in the case of MalDozer
dataset (Table 7.11), F1-Score of 85%. Our interpretation of this result comes from
Tables 7.12, 7.13, and 7.14, which depict the detailed results per malware family.
For example, the family agent exhibits poor results because of the mislabeling,
since agent is a common name for many Android malware families. We believe
that there is a lot of noise in family labeling of the MalDozer dataset since
we leverage only one security vendor for labeling. Despite this fact, MalDozer
demonstrates acceptable results and robustness.

Table 7.10 Attribution on Drebin

	F1%	P%	R%
2-Fold	98.1192	98.1401	98.1334
3-Fold	98.6882	98.6998	98.6912
5-Fold	98.5824	98.5961	98.5839
10-Fold	**98.5198**	**98.5295**	**98.5196**

The bold values represent the best values

Table 7.11 Attribution on MalDozer

	F1%	P%	R%
2-Fold	89.3331	89.5044	89.3424
3-Fold	81.8742	82.7565	81.8109
5-Fold	83.8518	84.1360	84.0061
10-Fold	**85.5233**	**85.6184**	**85.8479**

The bold values represent the best values

7.4.4   Runtime Performance

In this section, we evaluate the efficiency of **MalDozer**, i.e., the runtime during the deployment phase. We divide the runtime into two parts: (1) *Preprocessing time*: the required time to extract and preprocess the sequences of Android API method calls. (2) *Detection time*: time needed to predict a given sequence of API method calls. Here, we ask the following two questions: (1) *Which part in the preprocessing needs optimization?* (2) *Does the preprocessing time depend on the size of APK or DEX file?* To answer these questions, we randomly select 1K benign apps and 1K malware apps. We measure the preprocessing time and correlate it with the size of APK and DEX files. Figure 7.8 shows the experimentation results in the case of an IoT device [8]. The scattered charts depict the preprocessing time along with the size of the APK or DEX file for the mixed, the benign-only, and the malware-only datasets. From Fig. 7.8, it is clear that the preprocessing time is linearly related to the size of the DEX file. We perform the same experiment on a server and a laptop, and we get very similar results, as shown in Figs. 7.9 and 7.10. Finally, we notice that the size of benign apps tends to be bigger than the one of malicious apps. Thus, the preprocessing time of benign apps is longer. We measure the detection time according to the model complexity of different hardware. Figure 7.11a depicts the average preprocessing time, along with its standard deviation, related to each hardware. The server machines and the laptop spend, on average, 1 s in the preprocessing time, which is quite acceptable for production. Also, as mentioned previously, we do not optimize the current preprocessing workflow. On an IoT device [8], the preprocessing takes, on average, about 4 s, which is more than acceptable for such a small device. Figure 7.11b presents the detection time on average that is related to each hardware. First, it is noticeable that the standard deviation is quite negligible, i.e., the detection time is constant for all apps. Also, the detection time is very low for all the devices. As for the IoT device, the detection time is only 1.3 s. Therefore, the average time that **MalDozer** needs to decide for a

Table 7.12 MalDozer
android malware dataset

	Malware family	#Sample	F1-score
01	FakeInst	4822	96.15%
02	Dowgin	2248	84.24%
03	SmsPay	1544	81.61%
04	Adwo	1495	87.79%
05	SMSSend	1088	81.48%
06	Wapsx	833	78.85%
07	Plankton	817	94.18%
08	Agent	778	51.45%
09	SMSReg	687	80.61%
10	GingerMaster	533	76.39%
11	Kuguo	448	78.28%
12	HiddenAds	426	84.20%
13	Utchi	397	93.99%
14	Youmi	355	72.39%
15	Iop	344	93.09%
16	BaseBridge	341	90.50%
17	DroidKungFu	314	85.85%
18	SmsSpy	279	85.05%
19	FakeApp	278	93.99%
20	InfoStealer	253	82.82%
21	Kmin	222	91.03%
22	HiddenApp	214	76.71%
23	AppQuanta	202	99.26%
24	Dropper	195	77.11%
25	MobilePay	144	78.74%
26	FakeDoc	140	96.38%
27	Mseg	138	55.38%
28	SMSKey	130	81.03%
29	RATC	111	84.81%
30	Geinimi	106	95.58%
31	DDLight	104	90.55%
32	GingerBreak	103	84.87%

given app is 5.3 s on average in case of an IoT device, as we know that preprocessing takes most of the time (4/5.3).

7.4.4.1 Model Complexity Evaluation

In this section, we examine the effect of model complexity on the detection time. By model complexity, we mean the number of parameters in the model, as depicted in Table 7.15. Many hyper-parameters can influence the complex nature of the model, but we primarily consider the word2vec embedding size. The latter is crucial for the

Table 7.13 Malgenome
attribution dataset

	Malware family	#Sample	F1-score
01	DroidKungFu3	309	99.83%
02	AnserverBot	187	99.19%
03	BaseBridge	121	98.37%
04	DroidKungFu4	96	99.88%
05	Geinimi	69	97.81%
06	Pjapps	58	95.65%
07	KMin	52	99.99%
08	GoldDream	47	99.96%
09	DroidDreamLight	46	99.99%

Table 7.14 Drebin
attribution dataset

	Malware family	#Sample	F1-score
01	FakeInstaller	925	99.51%
02	DroidKungFu	666	98.79%
03	Plankton	625	99.11%
04	Opfake	613	99.34%
05	GinMaster	339	97.92%
06	BaseBridge	329	97.56%
07	Iconosys	152	99.02%
08	Kmin	147	99.31%
09	FakeDoc	132	99.24%
10	Geinimi	92	97.26%
11	Adrd	91	96.13%
12	DroidDream	81	98.13%
13	Glodream	69	90.14%
14	MobileTx	69	91.97%
15	ExploitLinuxLotoor	69	99.97%
16	FakeRun	61	95.16%
17	SendPay	59	99.14%
18	Gappusin	58	97.43%
19	Imlog	43	98.85%
20	SMSreg	41	92.30%

detection of the model, especially if we have a big dataset. Table 7.15 demonstrates
the complexity of the model versus the F1-Score. It is noticeable that the larger
the number of parameters is, the more its performance increases. Based on our
observation, bigger models are more accurate and more robust to changes. Finally,
Fig. 7.12 displays the execution time of the models in Table 7.15 on the IoT device.
The detailed execution related to all the hardware is presented in Fig. 7.12.

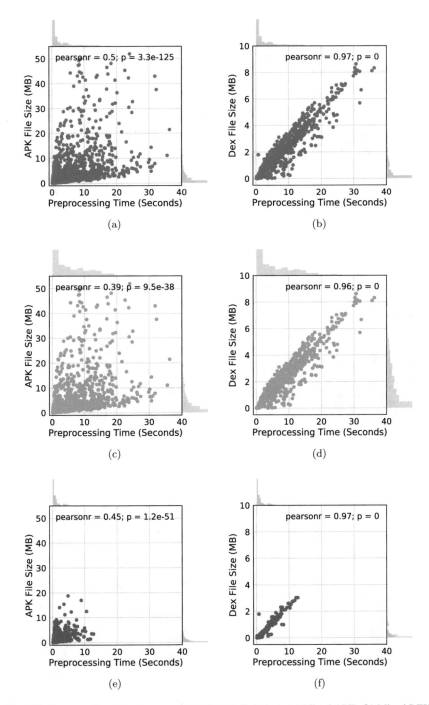

Fig. 7.8 Preprocessing time versus package sizes (IoT device). (**a**) Mixed APK. (**b**) Mixed DEX. (**c**) Benign APK. (**d**) Benign DEX. (**e**) Malware APK. (**f**) Malware DEX

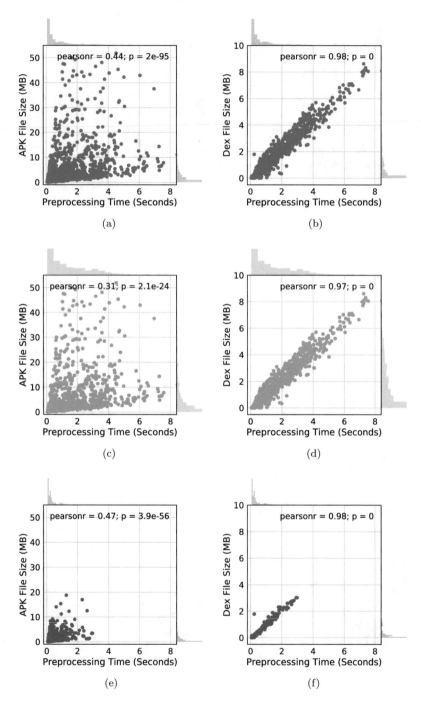

Fig. 7.9 Preprocessing time versus package sizes (laptop). (**a**) Mixed APK. (**b**) Mixed Dex. (**c**) Benign APK. (**d**) Benign Dex. (**e**) Malware APK. (**f**) Malware Dex

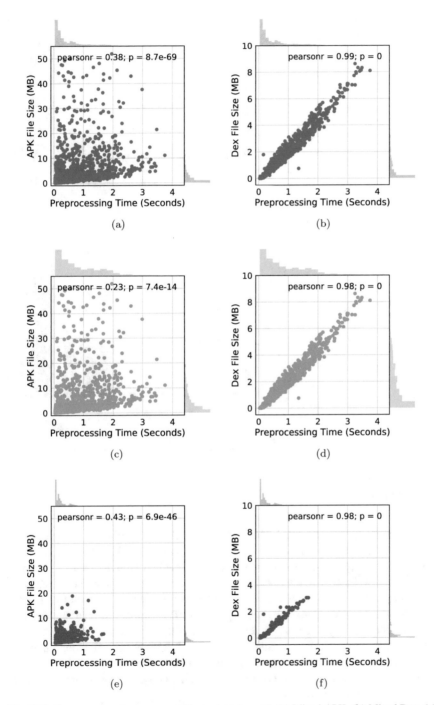

Fig. 7.10 Preprocessing time versus package sizes (server). (**a**) Mixed APK. (**b**) Mixed Dex. (**c**) Benign APK. (**d**) Benign Dex. (**e**) Malware APK. (**f**) Malware Dex

Fig. 7.11 Runtime versus
hardware. (**a**) Preprocess. (**b**)
Prediction

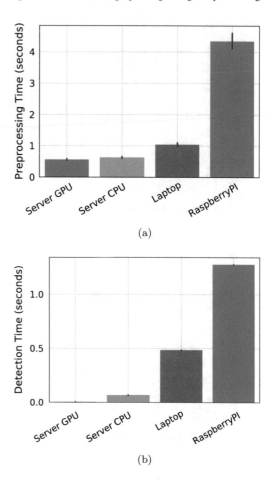

(a)

(b)

7.5 Summary

In this chapter, we presented MalDozer, an automatic, efficient, and effective
Android malware detection, and attribution system. MalDozer relies on deep
learning techniques and raw sequences of API method calls to identify Android
malware. We have evaluated MalDozer on several small and large datasets,
including *Malgenome*, *Drebin*, and our MalDozer dataset, in addition to a dataset
of benign apps downloaded from Google Play. The evaluation results show that
MalDozer is highly accurate in terms of a malware detection as well as their
attribution to corresponding families. Moreover, MalDozer can efficiently run under
multiple deployment architectures, ranging from servers to small IoT devices. This
work represents a step toward practical, automatic, and effective Android malware
detection and family attribution.

In the next chapter, we propose a system that focuses on the robustness and
the adaptability aspects in Android malware detection. First, the detection system

Table 7.15 Model complexity versus detection performance

	#Params	F1%	Word2Vec Size
Model 01	6.6 Million	98.95	100k
Model 02	4.6 Million	95.84	70k
Model 03	3.4 Million	93.81	50k
Model 04	1.5 Million	90.08	20k

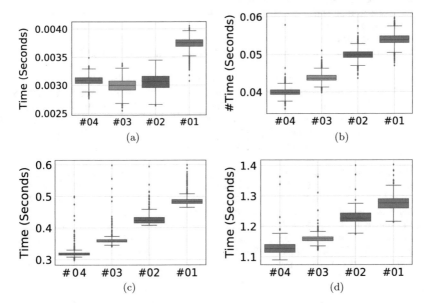

Fig. 7.12 Detection time versus model complexity. (**a**) Server GPU. (**b**) Server CPU. (**c**) Laptop. (**d**) IoT

should show resiliency to common code obfuscation and transformation techniques. Second, the detection performance should be resilient to the operating system, and malware changes overtime by employing an adaptation mechanism to handle changes.

References

1. T. Mikolov, I. Sutskever, K. Chen, G.S. Corrado, J. Dean, Distributed representations of words and phrases and their compositionality, in *Advances in Neural Information Processing Systems 26: 27th Annual Conference on Neural Information Processing Systems 2013. Proceedings of a Meeting Held December 5–8, 2013, Lake Tahoe, Nevada, United States* (2013), pp. 3111–3119
2. The Android Native Development Kit (NDK), https://developer.android.com/ndk/index.html. Accessed Jan 2016
3. Y. Kim, Convolutional neural networks for sentence classification, in *Proceedings of the 2014 Conference on Empirical Methods in Natural Language Processing, EMNLP 2014, October 25–29, 2014, Doha, Qatar, A Meeting of SIGDAT, a Special Interest Group of the ACL* (2014), pp. 1746–1751

4. J. Pennington, R. Socher, C.D. Manning, Glove: global vectors for word representation, in *Proceedings of the 2014 Conference on Empirical Methods in Natural Language Processing, EMNLP 2014, October 25–29, 2014, Doha, Qatar, A Meeting of SIGDAT, a Special Interest Group of the ACL* (2014), pp. 1532–1543
5. N. Srivastava, G.E. Hinton, A. Krizhevsky, I. Sutskever, R. Salakhutdinov, Dropout: a simple way to prevent neural networks from overfitting. J. Mach. Learn. Res. **15**(1), 1929–1958 (2014)
6. S. Ioffe, C. Szegedy, Batch normalization: accelerating deep network training by reducing internal covariate shift, in *Proceedings of the 32nd International Conference on Machine Learning, ICML 2015, Lille, France*, 6–11 July 2015, pp. 448–456
7. Tensorflow, https://www.tensorflow.org. Accessed Jan 2017
8. RaspberryPI 2, https://www.raspberrypi.org/products/raspberry-pi-2-model-b/. Accessed Jan 2017
9. Android Things, https://developer.android.com/things/. Accessed Sept 2016
10. Android Malware Genome Project, http://www.malgenomeproject.org/. Accessed Jan 2015
11. D. Arp, M. Spreitzenbarth, M. Hubner, H. Gascon, K. Rieck, DREBIN: effective and explainable detection of android malware in your pocket, in *21st Annual Network and Distributed System Security Symposium, NDSS 2014, San Diego, California, USA*, 23–26 Feb 2014
12. Contagiominidump Malware Repository, https://contagiominidump.blogspot.ca. Accessed Jan 2017
13. Playdrone Android Dataset, https://archive.org/details/playdrone-apks. Accessed Jan 2017
14. K. Allix, T.F. Bissyandé, J. Klein, Y.L. Traon, Androzoo: collecting millions of android apps for the research community, in *Proceedings of the 13th International Conference on Mining Software Repositories, MSR 2016, Austin, TX, USA*, 14–22 May 2016, pp. 468–471
15. E. Mariconti, L. Onwuzurike, P. Andriotis, E.D. Cristofaro, G.J. Ross, G. Stringhini, Mamadroid: detecting android malware by building Markov chains of behavioral models, in *24th Annual Network and Distributed System Security Symposium, NDSS 2017, San Diego, California, USA*, 26 Feb–1 March 2017

Chapter 8
Resilient and Adaptive Android Malware Fingerprinting and Detection

In this chapter, we present PetaDroid, an Android detection system that provides, in contrast to MalDozer (previous chapter), (1) resiliency to common obfuscation techniques by introducing code randomization during training; (2) adaptation to operating system and malware change overtime by introducing the use of confidence-based decisions to collect adaptation datasets overtime. In this context, we identify several limitations and gaps in the state-of-the-art Android malware detection solutions.

First, the accuracy of Android malware detection systems tends to decrease over time due to different factors (new OS versions, new malware families, new malicious techniques). We use time resiliency terminology to denote the adaptation of the Android malware detection solution to the change of malware, its attack techniques, and the Android platform. (1) It is important to note that detection systems may not be able to detect recently discovered samples because the system did not see such family samples during the training phase. However, the changes introduced by new families are, in most cases, incremental compared to existing malware threats. (2) Malware developers build attacks against Android devices exploiting these zero-day flaws. The developed malware samples may be part of an existing family, but such samples could also exhibit some variation in their implementation, depending on the employed exploitation techniques. Such variant malware indicates a progressive evolution that could deceive detection systems over time. (3) New APIs are introduced in the Android platform in every new version or update to the operating system and its ecosystem services. New APIs provide new capabilities for developers to build new app functionalities. On the other hand, malware developers could exploit these APIs to make new malicious functionalities to deceive existing Android malware detection. In time, the problem accumulates from 1 year to another, as seen in previous solutions [1, 2]. The gap between the detection capability and the emergence of new threats is indeed increasing, but the effect is minimum on a given time and incremental, which could be addressed timely. Therefore, we say that an Android malware detection solution is time-

E. B. Karbab et al., *Android Malware Detection Using Machine Learning*, Advances in Information Security 86, https://doi.org/10.1007/978-3-030-74664-3_8

resilient if it can adapt to the changes of the Android platform as well as the changes in benign and malicious app patterns. A crucial component to adaptation is the ability of the solution to generalize from a small number of samples. Therefore, the detection performance on a small training set is an essential requirement for modern solutions. Because the sooner our detection solution can grasp the malicious patterns from a small dataset, the faster we can detect such threat, especially at a market level.

Second, only a few existing solutions, [3, 4], provide Android malware family attribution functionality. Furthermore, these solutions are built using supervised learning where prior-knowledge on the families is required. However, such knowledge is hard to get and not realistic in many cases, especially for new families. Third, malware developers employ various obfuscation techniques to thwart detection attempts. Obfuscation resiliency is a key requirement in modern malware fingerprinting that employs static analyses. Very few solutions address the obfuscation issue in the context of Android malware detection. Existing obfuscation resilient solutions such as DroidSheive [5] require manual feature extraction. Fourth, solutions, such as DroidSheive [5], Stormdroid [6], and Drebin [3], rely on manual feature engineering based on classification techniques such as SVM and KNN. Despite their good detection performance, their approach is not scalable to the amount and the growth pace of Android malware. Therefore, there is an increasing need for solutions that are based on automatic feature engineering using deep learning and NLP techniques. Fifth, state-of-the-art solutions, such as MaMaDroid [1, 2], require a lot of computing power due to the complex preprocessing, which affects the overall efficiency. The last aspect is an important requirement for Android malware detection due to the growing number of Android apps.

PetaDroid aims at satisfying all the aforementioned issues/requirements of modern Android malware fingerprinting by addressing the previously identified gaps and challenges in existing state-of-the-art solutions.

8.1 Methodology

In this section, we detail PetaDroid methodology and its components.

8.1.1 Approach

PetaDroid employs static analysis techniques on Android Dalvik Virtual Machine (DVM) binary code (DEX) to check the maliciousness of Android apps. PetaDroid starts by extracting raw static features from the Android Packaging (APK), specifically the Dalvik VM bytecode (DEX). We develop a fast preprocessing phase to extract raw Dalvik assembly instructions. We generate, on the fly, the canonical form of the assembly instructions by substituting the value of constants, memory addresses with symbolic names. The output is a raw sequence of canonical assembly instructions of Dalvik virtual machine.

PetaDroid maintains a logical separation among the software component of applications. We keep track of classes and methods' instruction sequences within the application global instruction sequence. It is a natural breakdown because an Android app is a set of classes, and a class is a set of methods and attributes. The global app execution sequence is composed of a list of micro-execution paths (method sequences), through which the execution proceeds during runtime. The extracted canonical instruction sequences help preserving the underlying micro-execution paths of the app while without an emphasis on the global execution order. Micro-execution paths are instruction and API sequences of code functions (classes' methods).

Previous solutions [2] apply heavy and complex preprocessing to construct a global call graph to simulate runtime execution. In contrast, our extraction approach is lightweight (very little computation needed in the preprocessing) because we consider only the method order within each given class methods. We argue that tracking the method order is sufficient to identify malicious apps. It allows swift preprocessing in commodity hardware while maintaining the intended granularity. Furthermore, and in contrast with previous solutions [2], we adopt granular features using a canonical instruction followed by representation learning. We propose custom code modeling techniques for representation learning inspired by advanced natural language processing techniques. Specifically, we design and develop *Inst2Vec* and *InstNGram2Bag* code modeling techniques to model and discover latent information to produce embeddings from canonical instruction sequences.

In a nutshell, **PetaDroid** has six phases:

(1) **Representation Learning**: PetaDroid learns latent representations using unsupervised word2vec techniques [7].
(2) **Malware Detection**: PetaDroid employs neural network module as features' selector from the embedding representation. In the classification task, the training dataset guides feature selection to make the right detection outcome. PetaDroid classification system rests on ensemble deep learning models (CNN) that consume *Inst2Vec* embedding features to fingerprint malicious applications.
(3) **Detection Adaptation**: PetaDroid classification ensemble produces a detection confidence probability. Apps detected with high confidence, whether malicious or benign, will extend **PetaDroid** primary labeled dataset (used to build the current detection ensemble). Periodically, **PetaDroid** makes a new detection ensemble on the new dataset (primary and extended).
(4) **Code Representation**: For the detected malware, we produce feature vectors using N-grams bag of words and feature hashing techniques on top of the canonical instruction sequence. The outcome is what we call an InstNGram2Bag vector for each detected malware. An InstNGram2Bag vector summarizes the intrinsic semantic of Android malware.

(5) **Digest Generation**: We produce digests by applying deep neural auto-encoders [8] on the InstNGram2Bag vectors to produce a compact embedding or a digest for each malicious sample.

(6) **Malware Family Clustering**: PetaDroid clusters the flagged malicious apps into groups with high inter-similarity between their digests, and most likely of the same malware family. PetaDroid clustering system is based on DBScan[1] clustering algorithm.

8.1.2 Android App Representation

In this section, we present the preprocessing of Dalvik code and its representation into a sequence of canonical instructions. We seek the preservation of the maximum information about apps' behaviors while keeping the process very efficient. The preprocessing begins with the disassembly of an app bytecode to Dalvik assembly code, as depicted in Fig. 8.1.

We model the Dalvik VM assembly code as code fragments where each fragment is a method code in the Dalvik assembly. It is a natural separation because

```
// Object Creation
new-instance v10, java/util/HashMap
// Object Access
invoke-direct v10, java/util/HashMap
if-eqz v9, 003e
..
// Method Invocation
// * = Android/telephony
invoke-virtual v4, */TelephonyManager.getDeviceId()java/lang/String
move-result-object v11
// Method Invocation
invoke-virtual v4, */TelephonyManager.getSimSerialNumber()java/lang/String
move-result-object v13
// Method Invocation
invoke-virtual v4 */TelephonyManager.getLine1Number()java/lang/String
move-result-object v4
...
// Object Creation
new-instance v20, java/io/FileReader
const-string v21, "/proc/cpuinfo"
invoke-direct/range v20, v21, java/io/FileReader.init(java/lang/String)
new-instance v21, java/io/BufferedReader
...
move/from16 v2, v20
// Field Access
// * = Android/content/pm
iget-object v0, v0, */ApplicationInfo.metaData Android/os/Bundle
move-object/from16 v19, v0
```

Fig. 8.1 Android assembly from a malware sample

---

[1]https://en.wikipedia.org/wiki/DBSCAN.

Dalvik code D is composed of a set of classes $D = \{C_1, C_2, \ldots, C_s\}$. Each class C_i contains a set of methods $C = \{M_1, M_2, \ldots, M_k\}$ where we find actual assembly instructions. We preserve the order of Dalvik assembly instructions within methods while ignoring the global execution paths. Method execution is a possible *micro-behavior* for an Android app, while a global execution path is a likely *macro-behavior*. PetaDroid assembly preprocessing produces a list of instruction sequences $P = \{S_1, S_2, \ldots, S_h\}$ where each sequence S contains an ordered instruction $S = \langle I_1, I_2, \ldots, I_v \rangle$. Thus, a sequence S defines a possible micro-execution (or behavior) from the Android app's overall runtime execution.

As shown in Fig. 8.1, Dalvik assembly is too sparse. We want to keep the assembly instruction skeleton that reflects possible runtime behaviors with less sparsity. In PetaDroid (in contrast to MalDozer in previous chapter), we propose a canonical representation for Dalvik assembly code as shown in Fig. 8.2. The key idea is to keep track of the Android platform APIs and objects utilized inside the method assembly. In order to fingerprint malicious apps, the canonical representation will mostly preserve the actions and the manipulated system objects, such as sending SMS action or getting (setting) sensitive information objects. PetaDroid canonical representation covers three types of Dalvik assembly instructions namely *Method invocation*, *object manipulation*, and *field access*, as shown in Fig. 8.2. In the method invocation, we focus on the method call, *Package.ClassName.MethodName*, the parameters list, *Package.ClassName*, and the return type, *Package.ClassName*. In object manipulation, we capture the class object, *Package.ClassName*, that is being used. Finally, we track the access to system fields by capturing the field name, *Package.ClassName.FieldName*, and its type, *Package.ClassName*. Our manual inspections of Dalvik assembly for hundreds of malicious and benign samples shows that these three forms cover the essential of Dalvik assembly instructions.

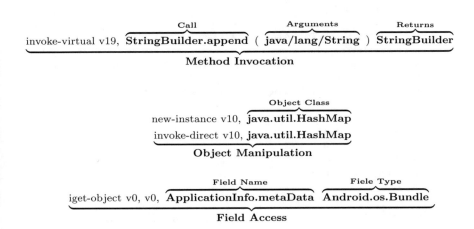

Fig. 8.2 Canonical representation of dalvik assembly

PetaDroid instruction parser keeps only the canonical representation and ignores the rest. For example, our experiments show that Dalvik opcodes add a lot of sparsity without enhancing the malware fingerprinting performance. On the contrary, it could affect the overall performance [9] negatively. The final step in the preprocessing of a method M (see Fig. 8.1) is to flatten the canonical representation of a method into a single sequence S (see Fig. 8.3). In the current design, we keep only Android platform related assets like API, classes, and system fields in the final method's sequence S. For this purpose, we maintain a vocabulary dictionary (assets names of Android platform) $V = \{\langle Asset_1, 1\rangle, \ldots, \langle Asset_d, d\rangle\}$ to filter and discretize the method sequence during the preprocessing. The output of the app representation phase is a list of sequences $P = \{cS_1, cS_2, \ldots, cS_h\}$. Each sequence is an ordered canonical instruction representation of one method.

8.1.3   Malware Detection

In this section, we present PetaDroid malware detection process using CNN on top of *Inst2Vec* embedding features. The detection process starts from a list of discretized canonical instruction sequences $P = \{cS_1, cS_2, \ldots, cS_h\}$. PetaDroid CNN ensemble produces a detection result together with maliciousness and benign detection probabilities for a given sample. In order to achieve automatic adaptation over time, we leverage the detection probabilities to automatically collect an extension dataset that PetaDroid employs to build new CNN ensemble models.

```
java/util/HashMap
java/util/HashMap
..
Android/telephony/TelephonyManager.getDeviceId()
java/lang/String
Android/telephony/TelephonyManager.getSimSerialNumber()
java/lang/String
Android/telephony/TelephonyManager.getLine1Number()
java/lang/String
...
java/io/FileReader
java/io/FileReader.init()
java/lang/String
java/io/BufferedReader
...
Android/content/pm/ApplicationInfo.metaData
Android/os/Bundle
```

Fig. 8.3 Flatten canonical representation from a malware sample

8.1.3.1 Fragment Detection

The fragment-based detection is a key technique in **PetaDroid** Android malware fingerprinting. A fragment F is a truncated portion from the concatenation cP of $P = \{cS_1, cS_2, \ldots, cS_h\}$. The size $|F|$ is the number of canonical instructions in the fragment F and it is a hyper-parameter in **PetaDroid** framework. For a sequence cS_i, the order of canonical instructions is preserved within a method. In other words, we guarantee the preservation of order inside the method sequence or what we refer to as a *micro-action*. On the other hand, no specific order is assumed between methods' sequences or what we refer to as *macro-action* (or behavior). In our context, and before we truncate cP into size $|F|$, we propose applying a random permutation on P to produce a random order in the macro-behavior while preserving its methods' micro-behaviors. The randomization happens in every access, whether it is during training or deployment phases. Each Android sample has $\frac{h!}{(h-k)!}$ possible permutation for the methods' sequences $P = \{cS_1, cS_2, \ldots, cS_h\}$, where h is the number of methods' sequence in a given Android app; and k is the number of sampled sequences. Notice that the size of the concatenated k sequences must be greater than $|F|$ (fragment size hyper-parameter).

The intuition behind fragment detection is the abstraction of Android apps behavior into a list of very small *micro-actions*. We consider each method canonical instruction sequence cS as possible *micro-actions* for an Android app. In a fragment, we keep the possible micro-actions intact and discard the app flow graph. We argue that this will force pattern learning, during the training, to focus on only micro-actions, which allows better generalization. Fragment-based detection has many advantages in the context of malware detection. First, fragment detection plays the role of dataset augmenter, which allows the learning model to generalize better from a small dataset. Second, it challenges the machine learning model and its training process to learn dynamic patterns at every training epoch. In other words, it focuses the model on robust, distinctive patterns from a sample of random micro-actions of methods. Third, we argue that our fragment-based detection helps improving the robustness of the malware detection model against conventional obfuscation techniques and code transformation in general. Fourth, in the testing phase, **PetaDroid** infers the maliciousness of a given sample by applying **PetaDroid** CNN on multiple sample fragments to obtain a detection decision with a specific confidence interval.

8.1.3.2 Inst2Vec Embedding

Inst2Vec is based on *word2vec* [7] technique to produce an embedding vector for each canonical instruction in our sequences. *Inst2Vec* is trained on instruction sequences to learn instructions semantics from the underlying contexts. This means that *Inst2Vec* learns a dense representation of a canonical instruction that reflect the instruction co-occurrence and context. The produced embeddings capture the semantics of instructions and translate into geometric values over multiple dimensions. Our Android malware detection technique is inspired by word2vec, modern NLP techniques as well as neural machine translation techniques. Furthermore, embedding features show high code fingerprinting accuracy and resiliency to common obfuscation techniques [10]. Word2vec [7] is a vector space model to represent the words of a document in a continuous vector space where words with similar semantics are mapped closely in the space. From a security perspective, we want to map our features (canonical instructions in a fragment) to continuous vectors where their semantics is translated to a distance in the vector space. Word2vec is a neural probabilistic model that is trained using the maximum likelihood concept. More precisely, given sequence of words: $w_1, w_2, \cdots, w_T$, at each position $t = 1, \cdots, T$, the model predicts a context of sequence within a window of fixed size m given center word w_j (illustrated in Eq. 8.1), where m is the size of the training context [7].

$$L(\theta) = \prod_{t=1}^{T} \prod_{-m \leq j \leq +m, j \neq 0} P(w_{t+j}|w_t; \theta) \tag{8.1}$$

The objective function [7] $J(\theta)$ is the negative log-likelihood as shown in Eq. 8.3. The probability $P(w_{t+j}|w_t; \theta)$ is defined in Eq. 8.5, where v_w and v'_w are the input and the output of the embeddings of w.

$$J(\theta) = -\frac{1}{T} \log L(\theta) \tag{8.2}$$

$$= -\frac{1}{T} \sum_{t=1}^{T} \sum_{-m \leq j \leq m, j \neq 0} \log P(w_{t+j}|w_t; \theta) \tag{8.3}$$

$$P(w_O|w_I) = \text{softmax}(\grave{v}_{w_O}^T v_{w_I}) \tag{8.4}$$

$$= \frac{\exp(\grave{v}_{w_O}^T v_{w_I})}{\sum_{w=1}^{W} \exp(\grave{v}_{w}^T v_{w_I})} \tag{8.5}$$

We train the embedding model by maximizing log-likelihood as illustrated in Eq. 8.7.

$$J_{\text{ML}} = \log P(w_O|w_I) \tag{8.6}$$

$$= (\grave{v}_{w_O}^T v_{w_I}) - \log \left(\sum_{w=1}^{W} \exp\{\grave{v}_{w}^T v_{w_I}\} \right) \tag{8.7}$$

8.1.3.3 Classification Model

Our single CNN model takes *Inst2Vec* features, which are a sequence of embeddings, each embedding captures the semantics of instructions. The temporal convolutional neural network [11], or 1-Dimensional CNN [12], is the working core component in **PetaDroid** single classification model. Table 8.1 details the architecture of our CNN single model.

The non-linearity used in our model employs the rectified linear unit (ReLUs) $h(x) = \max\{0, x\}$. We used Adam [13] optimization algorithm with a mini-batch of size 32 and a learning rate $3e - 4$ for 100 epochs in all our experiments.

8.1.3.4 Dataset Notation

In this section, we present the notations that will be used in the next sections.

$X = \{(\langle cP_0, y_0 \rangle, \langle cP_1, y_1 \rangle, .., \langle cP_m, y_m \rangle\}$: X is the global dataset used to build ensemble models and report **PetaDroid** performance on various tasks. Where m is number of $\langle sample, label \rangle$ records in the global dataset X.

$X = \{X_{build}, X_{test}\}$: We use a build set X_{build} to train and tune the hyperparameters of **PetaDroid** models. The test set X_{test} represents Android apps that the system will receive during the deployment. The test set X_{test} is used to measure the final performance of PetaDroid, which is reported in the evaluation section. X is split randomly into X_{build} (50%) and X_{test} (50%).

$X_{build} = \{X_{train}, X_{valid}\}$: The build set, X_{build}, is composed of a training set X_{train} and a validation set X_{valid}. It is used to build **PetaDroid** single CNN models for the CNN ensemble. For each single CNN model, we tune the model parameters to achieve the best detection performance on X_{valid}. The build set $m_{build} = m_{train} + m_{valid}$: is the total number of records used to build **PetaDroid**. The training set takes 80% of the build set X_{build}, and 20% of X_{build} is used for the validation set X_{valid}.

Table 8.1 PetaDroid CNN detection model

#	Layers	Options
1	1D-Conv	Filter=128, Kernel=(5,5), Stride=(1,1), Padding=0, Activation=ReLU
2	BNorm	BatchNormalization
3	Global max pooling	/
4	Linear	#Output=512 , Activation=ReLU
5	Linear	#Output=256 , Activation=ReLU
6	Linear	#Output=1 , Activation=ReLU

8.1.3.5   Detection Ensemble

PetaDroid detection component relies on an ensemble $\Phi = \langle sC_1, sC_2, \ldots, sC_\phi \rangle$. Ensemble Φ is composed of ϕ single CNN models. The number of single CNN models in the ensemble ϕ is a hyper-parameter. We fixed $\phi = 6$ in the evaluation experiments. As mentioned previously **PetaDroid** trains each CNN model C for number of epochs (*epochs* = 100). In each epoch, we compute $Loss_T$ and $Loss_V$, the *training* and *validation* losses, respectively, and save a snapshot of the single CNN model parameters. $Loss_T$ and $Loss_V$ are the log loss across training and validation sets:

$$p = singleCNN_\theta(y = 1|cP)$$

$$loss(y, p) = -(y \log(p) + (1 - y) \log(1 - p)),$$

$$Loss_T = \frac{-1}{m_{train}} \sum_{i=1}^{m_{train}} loss(y_i, p_i),$$

$$Loss_V = \frac{-1}{m_{valid}} \sum_{i=1}^{m_{valid}} loss(y_i, p_i)$$

where p is the maliciousness likelihood probability given a fragment F (a concatenated and truncated canonical instructions cP) and model parameters θ (Sect. 8.1.2). **PetaDroid** selects automatically the top ϕ models from the saved model snapshots that have the lowest *training* and *validation* losses $Loss_T$ and $Loss_R$, respectively.

$$\hat{y} = \Phi(x) \quad = \frac{1}{\phi} \left(\sum_{i}^{\phi} sC_i(x) \right) \tag{8.8}$$

   PetaDroid CNN ensemble Φ produces a maliciousness probability likelihood by averaging the likelihood probabilities of single CNN models sC, as shown in Eq. 8.8.

8.1.3.6   Confidence Analysis

PetaDroid ensemble computes the maliciousness probability likelihood $Prob_{Mal}$ given a fragment F, as follows:

$$\hat{y} = \Phi(F), \quad Prob_{Mal} = \hat{y}, \quad Prob_{Ben} = (1 - \hat{y})$$

Previous Android malware detection solutions, such as [1, 4, 6], utilize a simple detection technique (we refer to it as a *general decision*) to decide on the maliciousness of Android apps. In the *general decision*, we compute general threshold

$\zeta \in [0, 1]$ that achieves the highest detection performance on the validation dataset X_{valid}. In the deployment phase (or evaluation in our case on X_{test}), the general decision D_ζ utilize the computed threshold ζ to make detection decisions:

$$D_\zeta = \begin{cases} Malware & Prob_{Mal} > \zeta \\ Benign & Prob_{Mal} <= \zeta \end{cases}$$

PetaDroid employs F1-score as detection performance metric to automatically select ζ and to report general detection performance on the test set X_{test} during our evaluation, in Sect. 8.2. We choose F1-score as our detection performance metric due to its simplicity, and its measurement reflects the reality under unbalanced datasets like in our case. Besides F1-score, we could use other metrics like ROC, precision, or recall. The general decision strategy is simple and effective in system development. It provides a firm decision for every sample. On the other hand, the security practitioner might prefer dealing with decisions that have associated confidence values and filter out less-confident decisions for further investigation. In a real deployment, we would like to have as many as possible detection decisions with high confidence and filter out the few uncertain apps that have low confidence probability. Unfortunately, the general decision strategy does not provide such functionality. For this purpose, we propose the **confidence decision strategy**, a mechanism to automatically filter out apps with uncertain decisions. PetaDroid computes a confidence threshold η that achieves not only a high detection performance (F1-score) but also a negligible error rate (false negative and false-positive rates) in the validation dataset. In other words, we add the error rate constraint to the system that computes the detection threshold η from X_{valid}. In the deployment, we make confidence-based decision as follows:

$$D_\eta = \begin{cases} Uncentain & Prob_{Mal} < \eta \wedge Prob_{Ben} < \eta \\ Malware & Prob_{Mal} >= \eta \wedge Prob_{Mal} > Prob_{Ben} \\ Benign & Prob_{Ben} >= \eta \wedge Prob_{Ben} > Prob_{Mal} \end{cases}$$

For example, we could fix the error rate to $< 1\%$ and automatically find η that achieves the highest F1-score in the validation set. Our goal is to maximize certain detection decisions on the deployment, which we called the *detection coverage performance* and minimize alerts for uncertain ones that require further analyses and investigation. In our case, the *detection coverage performance* is the percentage of confidence decisions from X_{test}. In Sect. 8.2, we conduct experiments where we report *general detection performance* metric in order to compare with existing solutions such as [1, 4, 6]. In addition, we report *confidence detection performance* and *detection coverage performance* metrics which we believe are more suitable for real-world deployment. Furthermore, the *confidence decision strategy* is key in PetaDroid retraining process, aiming toward automatic adaptation as will be explained next.

8.1.3.7   PetaDroid Adaptation Mechanism

In this section, we describe our mechanism to adapt to Android ecosystem changes overtime automatically. The key idea is to retrain the CNN ensemble on new benign and malware samples at every epoch to learn the latest changes. To enhance the automatic adaptation, we leverage the confidence analysis to collect an extension dataset that captures the incremental change over time. Initially, we train **PetaDroid** ensemble using $X_{build} = \{X_{train} + X_{valid}\}$. Afterward, **PetaDroid** leverages the *confidence detection strategy* to build an extension dataset X_{exten} from test dataset X_{test} with high-confidence detected apps. In a real deployment, X_{test} is a stream of Android apps that needs to be checked for maliciousness by the vetting system. The test dataset $X_{test} = \{X_{Certain}, X_{Uncertain}\}$ is composed of apps having a high-confidence decision ($X_{Certain}$ or X_{exten}) and apps having uncertain decisions $X_{Uncertain}$. In the deployment, **PetaDroid** accumulates high-confidence apps over time to form X_{exten} dataset. At every time epoch, **PetaDroid** utilizes the extension dataset X_{exten} to extend the original X_{build} and later updates the CNN ensemble models. In our evaluation, and after updating the CNN ensemble, we report **updated general performance** and **updated confidence-based performance**, which are respectively the general and confidence-based performance of the new trained CNN ensemble on X_{test}. They answer the question: what would be the detection performance on $X_{test} = \{X_{Certain}, X_{Uncertain}\}$ after we build the ensemble on $X_{NewBuild} = \{X_{Certain}, X_{build}\}$? In other words, **PetaDroid** reviews previous detection decisions using the new CNN ensemble and drives new general and confidence-based performance.

In a deployment environment, **PetaDroid** is continuously receiving new Android apps, whether benign or malware, represented by X_{test} in our evaluation. **PetaDroid** employs the extension dataset X_{exten} to overcome pattern changes, whether malicious or benign, automatically. Our approach is based on the assumption that Android apps patterns change incrementally with slow progress. Therefore, starting from a relatively small X_{build} dataset, **PetaDroid** could learn new patterns from new X_{exten} dataset progressively over time. **PetaDroid** ensemble update is an automatic operation for every period. During offline analyses, **PetaDroid** data extension process could be employed to improve the classification result on a fixed test dataset X_{test} starting from a small X_{build}. Our evaluation (Sect. 8.2.6) shows the effectiveness of our update strategy.

8.1.4   Malware Clustering

In this section, we detail the family clustering system of **PetaDroid**. **PetaDroid** clustering aims at grouping the previously detected malicious apps (Sect. 8.1.3) into highly similar groups of malicious apps, which are most likely part of the same malware family. **PetaDroid** clustering process starts from a list of discretized canonical instruction sequences $P = \langle cS_1, cS_2, \ldots, cS_h \rangle$ of the detected malicious apps. We introduce the *InstNGram2Vec* technique and deep neural network auto-encoder to generate embedding digests for malicious apps. Afterward, we cluster malware digests using the DBScan clustering algorithm to generate malware family groups.

8.1.4.1 InstNGram2Vec

InstNGram2Vec is a technique that maps concatenated instruction sequences to fixed-size embeddings employing NLP bag of words N-grams [14] and feature hashing [15] techniques.

Common N-Gram Analysis (CNG)

The common N-gram analysis (CNG) [14], or simply N-gram, has been extensively used in text analyses and natural language processing in general and related applications such as automatic text classification and authorship attribution [14]. N-gram computes the contiguous sequences of n items from a large sequence. In the context of **PetaDroid**, we compute canonical instructions N-grams on concatenated sequence cP by counting the instruction sequences of size n. Notice that the N-grams are extracted using a forward moving window (of size n) by one step and incrementing the counter of the found features (instruction sequence in the window) by one. The window size n is a hyper-parameter; we fixed $n = 4$ in all our experimentations. N-gram computation takes place simultaneously with the feature hashing in the form of a pipeline to prevent and limit computation and memory overuse due to the high dimensionality of N-grams.

Feature Hashing.

PetaDroid employs Feature Hashing (FH) [15] along with N-grams to vectorize cP. The feature hashing algorithm takes as an input cP N-grams generator and the target length L of the feature vector. The output is a feature vector with components x_i and a fixed size L. In our framework, we fix $L = |V|$, where V is the vocabulary dictionary (Sect. 8.1.2). We normalize x_i using the Euclidean norm (also called L2 norm). Applying *InstNGram2Vec* on a detected malicious app cP produces a fixed-size hashing vector hv. Therefore, the result is $HV = \{hv_0, hv_1, \dots, hv_{DMal}\}$, and hashing vector hv for $DMal$ detected malicious apps.

8.1.4.2 Deep Neural Auto-Encoder and Digest Generation

We develop a deep neural auto-encoder through stacked hidden layers of encoding and decoding operations, as shown in Table 8.2. The proposed auto-encoder learns the latent representation of Android apps in an unsupervised way. The unsupervised learning of the auto-encoder is done through the reconstruction (Table 8.2) of the unlabeled hashing vectors $HV = \{hv_0, hv_1, \ldots, hv_{DMal}\}$ of random Android apps. Notice that we do not need any labeling during the training of **PetaDroid** auto-encoder, off-the-self Android apps are sufficient.

The training goal is to make the auto-encoder learn to efficiently produce a latent representation (or digest) of an Android app hv that keeps the discriminative patterns of malicious and benign Android apps. Formally, the input to the deep neural auto-encoder [8] network is an unlabeled hash vector $HV = \{hv_0, hv_1, \ldots, hv_{DMal}\}$, denoted $x' \in \mathcal{U}$ on which operates the encoder network $f_{enc} : \mathbb{R}^{|V|} \to \mathbb{R}^p$, $p = 64$ as shown in Table 8.2 (parameterized by Θ_{enc}) to produce the latent representation $z_{x',\Theta_{enc}}$, i.e.

$$z_{x',\Theta_{enc}} = f_{enc}(x'; \Theta_{enc}) \tag{8.9}$$

The produced digest, namely $z_{x',\Theta_{enc}} \in \mathbb{R}^p$, is used by the decoder network $f_{dec} : \mathbb{R}^p \to \mathbb{R}^{|V|}$ to rebuild or reconstruct the InstNGramBag2Vec feature vector. The training loss of the auto-encoder network given the unlabeled hash vector x' is,

$$\tilde{x}' = f_{dec}(z; \Theta_{dec}) \tag{8.10}$$

$\tilde{x}' \in \mathbb{R}^{d \times w}$ denotes the generated reconstruction.

$$\mathcal{L}_{auto}(x'; \Theta_{enc}, \Theta_{dec}) = \|x' - f_{dec}(z_{x',\Theta_{enc}}; \Theta_{dec})\|^2 \tag{8.11}$$

In the training phase, the gradient-based optimizer minimizes the objective reconstruction function on the InstNGramBag2Vec feature vectors of unlabeled Android apps.

Table 8.2 Architecture PetaDroid deep neural auto-encoder

#	Layers	Options
01	Linear	#Output=$|V|$, Activation=Tanh
02	Linear	#Output=512, Activation=Tanh
03	Linear	#Output=256, Activation=Tanh
04	Linear	#Output=128, Activation=Tanh
05	Linear	#Output=64, Activation=Tanh
06	Linear	#Output=64, Activation=Tanh
07	Linear	#Output=128, Activation=Tanh
08	Linear	#Output=256, Activation=Tanh
08	Linear	#Output=512, Activation=Tanh
10	Linear	#Output=$|V|$, Activation=Tanh

$$(\Theta_{enc}^*, \Theta_{dec}^*) = \arg \min_{\Theta_{enc}, \Theta_{dec}} \sum_{i=1}^{N_1+N_2} \mathscr{L}_{auto}(x'_i; \Theta_{enc}, \Theta_{dec}) \qquad (8.12)$$

Notice that PetaDroid auto-encoder is trained only once during all the experimentation due to its general usage. To this end, **PetaDroid** employs a trained encoder f_{dec} to produce digests $Z = \{z_0, z_1, \ldots, z_{DMal}\}$ for the detected malicious apps.

8.1.4.3 Malware Family Clustering

PetaDroid clusters the detected malware digests $Z = \{z_0, z_1, \ldots, z_{DMal}\}$ into groups of malware with high similarity and most likely belonging the same family. In **PetaDroid** clustering, we use an **exclusive** clustering mechanism. This means that we do not cluster all the detected malicious apps. The clustering algorithm only groups highly similar samples and tags the rest as unclustered. This feature is convenient in real-world deployments since we might not always detect malicious apps from the same family, and we would like to have family groups only if there are groups of the sample malware family. To achieve this feature, we employ the *DBScan* clustering algorithm. *DBScan*, in contrast with clustering algorithms such as *K-means*, produces clusters with high confidence. The most important metrics in **PetaDroid** clustering is the homogeneity of the produced clusters.

8.1.5 Implementation

We build **PetaDroid** using `Python` and `Bash` programming languages. We use `dexdump`[2] to disassemble Android app DEX code into Dalvik assembly. The tool `dexdump` is a simple, yet very efficient tool to parse APK file and produce disassembly in a textual form. We develop python and bash scripts to parse Dalvik assembly to produce sequences of canonical instructions. Notice that there is no optimization in the preprocessing; in the efficiency evaluation, we only use a single thread script for a given Android app. We implement **PetaDroid** neural networks, CNN ensemble, and auto-encoders, using PyTorch.[3] For clustering, we employ official `hdbscan`[4] implementation. We evaluate the efficiency of **PetaDroid** on a commodity hardware server (Intel(R) Xeon(R) CPU E5-2630, 2.6GHz). For training, we use *NVIDIA TitanX* Graphic Processing Unit (GPU).

---

[2]https://tinyurl.com/y4ze8nyy.

[3]https://pytorch.org.

[4]https://en.wikipedia.org/wiki/DBSCAN.

8.2   Evaluation

In this section, we evaluate **PetaDroid** framework through a set of experiments and settings involving different datasets. We aim to answer questions such as: *What is the detection performance of* **PetaDroid** *on datasets with various sizes (Sect. 8.2.2)? What is the effect of* **PetaDroid** *ensemble and build dataset sizes on the overall performance (Sect. 8.2.2.2)? What is the performance of family clustering (Sect. 8.2.3)? How efficient is* **PetaDroid** *in terms of runtime on commodity machines (Sect. 8.2.7)? How robust is PetaDroid against common obfuscation techniques (Sect. 8.2.4)?*

8.2.1   Android Dataset

Our evaluation dataset contains 9.7 million Android apps (the dataset size is 100 Tera bytes) collected across the last 10 years from August 2010 to August 2019, as depicted in Table 8.3. The extensive coverage in size (9.7 M), time range (06-2010 to 08-2019), and malware families (+300 family) make the result of our evaluation quite compelling. First, we leverage reference Android malware datasets namely MalGenome [16], Drebin [3], MalDozer [4], and AMD [17]. We use these datasets in the evaluation of both **PetaDroid** detection and family clustering because they have family labels. A reference dataset helps comparing our evaluation results with the related work. Also, we collected Android malware from VirusShare[5] malware repository. This dataset serves in the evaluation of **PetaDroid** detection. For benign apps, we randomly select samples from the AndroZoo [18] dataset (7.4 Million benign samples), which are two to seven times the size of reference malware dataset in each experiment.

Table 8.3 Evaluation datasets

Name	#samples	#families	Time
MalGenome [16]	1.3K	49	2010–2011
Drebin [3]	5.5k	179	2010–2012
MalDozer [4]	21k	20	2010–2016
AMD [17]	25k	71	2010–2016
VirusShare[a]	33k	/	2010–2017
MaMaDroid [2]	40k	/	2010–2017
AndroZoo [18]	9.5M	/	2010–Aug 2019

[a] https://VirusShare.com

---

[5] https://VirusShare.com.

In the comparison between PetaDroid, MaMaDroid [1, 2], and DroidAPIMiner [19], we apply PetaDroid on the same dataset (benign and malware) used in MaMaDroid evaluation[6] to measure the performance of PetaDroid against state-of-the-art Android malware detection solutions.

In our use cases, we employ the whole AndroZoo[7] [18] dataset (the collection ends on August 2019), which contains 7.4 million benign apps and 2.1 million malware apps. We rely on VirusTotal engine that utilizes multiple anti-malware vendors in (metadata provided by AndroZoo repository) to label the samples. As shown in Fig. 8.4, the dataset covers more than 10 years. To assess PetaDroid obfuscation resiliency, we conduct an obfuscation evaluation on PRAGuard dataset,[8] which contains 11k obfuscated malicious apps using common obfuscation techniques [20]. In addition, we generate over 100k benign and malware obfuscated Android apps employing DroidChameleon obfuscation tool [21] using common obfuscation techniques and their combinations.

8.2.2 Malware Detection

In this section, we report the detection performance of PetaDroid and the effect of hyper-parameters on malware detection performance.

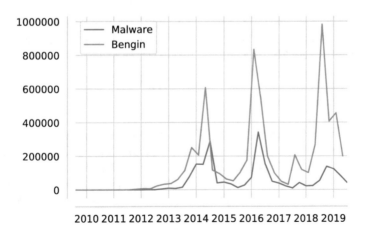

Fig. 8.4 AndroZoo benign and malware distribution over time

---

[6]https://bitbucket.org/gianluca_students/mamadroid_code/src/master/.

[7]https://androzoo.uni.lu/.

[8]http://pralab.diee.unica.it/en/AndroidPRAGuardDataset.

8.2.2.1 Detection Performance

Table 8.4 shows **PetaDroid** *general* and *confidence-based* performance in terms of F1-score, recall, and precision metrics on the reference datasets. In the general performance, **PetaDroid** achieves high F1-score 96–99% with low false-positive rate (recall score of 95.7–99.5%). The detection performance is higher under confidence settings. The F1-score is 99% and very low false-positive rate with a recall score of 99.8% on average. The confidence-based performance causes the filtration of 1–8% low confidence samples from the testing set. In all our experiments, the confidence performance flags ≈6% on average, as uncertain decisions, which is a small and realistic value in a deployment with low false positives. The filtered Android apps are flagged as suspicious apps, which might need further attention from the security practitioner.

8.2.2.2 Dataset Size Effect

One of the advantages of fragment-based malware detection is the data augmentation of the building dataset by random shuffles. **PetaDroid**, as shown in Table 8.5, exploits this feature to enhance the detection performance on small build datasets. In Table 8.5, there is a small change in the detection performance when the build set percentage drops from 90 to 50% from the overall dataset. Note that the build dataset is already composed of 80% training and 20% validation set $X_{build} = \{X_{train}, X_{valid}\}$, which makes the model trained on a smaller dataset. However, **PetaDroid** detection still performs well under these settings. Notice that in all our experiments, we use 50% from the evaluation dataset as a build dataset.

8.2.2.3 Ensemble Size Effect

Another important factor that affects **PetaDroid** malware detection performance is the number of CNN models in the detection ensemble. Table 8.6 depicts **PetaDroid** performance under different ensemble sizes. We notice the high detection accuracy using single CNN model (95–99% F1-score). In addition to the strength of CNN in discriminating Android malware, fragment detection adds a significant value to

Table 8.4 General and confidence performances on various reference datasets

	General (%)			Confidence (%)		
Name	F1	P	R	F1	P	R
Genome	99.1	99.5	98.6	99.5	100	99.0
Drebin	99.1	99.0	99.2	99.6	99.6	99.7
MalDozer	98.6	99.0	98.2	99.5	99.7	99.4
AMD	99.5	99.5	99.5	99.8	99.7	99.8
VShare	96.1	96.4	95.7	99.1	99.7	98.6

Table 8.5 Effect of building dataset size on the detection performance

Build dataset size (%)	General (F1 %)			Confidence (F1 %)		
	50%	70%	90%	50%	70%	90%
Genome	98.8	99.1	98.8	100	99.5	99.1
Drebin	98.2	99.1	99.1	99.6	99.6	99.8
MalDozer	98.3	98.6	98.7	99.6	99.5	99.6
AMD	99.3	99.5	99.5	99.7	99.8	99.7
VShare	95.6	96.1	96.4	99.0	99.1	99.1

Table 8.6 Effect of ensemble size on detection

#model	General (F1 %)				Confidence (F1 %)			
	1	5	10	20	1	5	10	20
MalGenome	99.5	99.3	99.3	99.1	99.5	99.5	99.5	99.5
Drebin	99.0	99.1	99.0	99.1	99.4	99.6	99.6	99.6
MalDozer	98.0	98.6	98.4	98.6	99.0	99.5	99.5	99.5
AMD	99.3	99.5	99.5	99.5	99.5	99.8	99.7	99.8
VShare	95.0	96.0	96.1	96.1	98.1	99.0	99.2	99.1

the overall performance even in a single CNN mode. In the case of MalGenome (Table 8.6), the ensemble size adds no value to the detection performance due to small size of MalGenome dataset ($1.3k$ malware + $12k$ benign randomly sampled from AndroZoo [18]). In case of VirusShare (Table 8.6), augmenting the ensemble size enhanced the detection rate. Our empirical tests show that $\phi = 6$ as the ensemble size gives good detection results. Increasing beyond $\phi = 6$ will have negligible benefits.

8.2.3 Family Clustering

In this section, we present the results of **PetaDroid** family clustering on reference datasets. Malware family clustering phase comes after **PetaDroid** detects a considerable number of malicious Android apps. The number of detected apps could vary from $1k$ (MalGenome [16]) to $24k$ (AMD [17]) samples depending on the deployment. We use *homogeneity* [22] and *coverage* metrics to measure the family clustering performance. The homogeneity metric scores the purity of the produced family clusters. A perfect homogeneity means each produced cluster contains samples from only one malware family. **PetaDroid** clustering aims only to generate groups with confidence-based while ignoring less certain groups. The coverage metrics score the percentage of the clustered dataset with confidence.

Figure 8.5 summarizes the clustering performance in terms of *homogeneity* and *coverage* scores. **PetaDroid** can produce clusters with high *homogeneity* 90–96% while keeping an acceptable *coverage*, 50% on average. At first glance, 50%

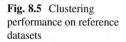

Fig. 8.5 Clustering performance on reference datasets

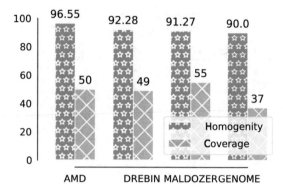

coverage seems to be a modest result, but we argue that it is satisfactory because: (1) we could extend the coverage, but this might affect the quality of the produced clusters. In the deployment, high-confidence clusters with minimum errors and acceptable coverage might be better than perfect coverage (in case of K-means clustering algorithm) with a high error rate. (2) The evaluation datasets have long tail malware families, meaning that most families have only a few samples. This makes the clustering very difficult due to the few samples (less than five samples) in each malware family in the detected dataset. In a real deployment, we could add unclustered samples to the next clustering iterations. In this case, we might accumulate enough samples to cluster for the long tail malware families.

8.2.4   Obfuscation Resiliency

In this section, we report **PetaDroid** detection performance on obfuscated Android apps. We experiment on: (1) PRAGuard obfuscation dataset [20] ($10k$) and (2) obfuscation dataset generated using DroidChameleon [21] obfuscation tool ($100k$). In the PRAGuard experiment, we combine Praguard dataset with $20k$ benign Android apps randomly sampled from benign app of AndroZoo repository. We split the dataset equally into build dataset $X_{build} = \{X_{train}, X_{valid}\}$ and test dataset X_{test}. Table 8.7 presents the detection performance of **PetaDroid** on different obfuscation techniques. **PetaDroid** shows high resiliency to common obfuscation techniques by having almost perfect detection rate, 99.5% F1-score on average.

In the DroidChameleon experiment, we evaluate **PetaDroid** on other obfuscation techniques, as shown in Table 8.8. The generated dataset contains obfuscated benign (apps originally from AndroZoo) and malware samples (originally from Drebin). In the building process of CNN ensemble, we only train with one obfuscation technique (Table 8.8) and make the evaluation on the rest of the obfuscation techniques. Table 8.8 reports the result of obfuscation resiliency on DroidChameleon generated dataset. The results show the robustness of PetaDroid. According to this experiment, **PetaDroid** is able to detect malware obfuscated with

Table 8.7 PetaDroid obfuscation resiliency on PRAGuard dataset

		General performance (%)		
ID	Obfuscation techniques	F1 (%)	P (%)	R (%)
1	Trivial	99.4	99.4	99.4
2	String encryption	99.4	99.3	99.4
3	Reflection	99.5	99.5	99.5
4	Class encryption	99.4	99.4	99.5
5	(1) + (2)	99.4	99.4	99.4
6	(1) + (2) + (3)	99.4	99.3	99.5
7	(1) + (2) + (3) + (4)	99.5	99.4	99.6
	Overall	99.5	99.6	99.4

Table 8.8 PetaDroid obfuscation resiliency on droidchameleon generated dataset

	General performance		
Obfuscation techniques	F1 (%)	P (%)	R (%)
No obfuscation	99.7	99.8	99.6
Class renaming	99.6	99.6	99.5
Method renaming	99.7	99.7	99.7
Field renaming	99.7	99.8	99.7
String encryption	99.8	99.8	99.7
Array encryption	99.8	99.8	99.7
Call indirection	99.8	99.8	99.7
Code reordering	99.8	99.8	99.7
Junk code insertion	99.8	99.8	99.7
Instruction insertion	99.7	99.8	99.7
Debug information removing	99.8	99.8	99.7
Disassembling and reassembling	99.8	99.8	99.7

common techniques even if the training is done on non-obfuscated datasets. We
believe that **PetaDroid** obfuscation resiliency comes from the usage of (1) Android
API (canonical instructions) sequences as features in the machine learning development. Android APIs are crucial in any Android app. A malware Developer cannot
hide API access, for example, *SendSMS* unless the malicious payload is downloaded
at runtime. Therefore, **PetaDroid** is resilient to common obfuscations as long as
they do not remove or hide API access calls. (2) The other factor is fragment-
randomization, which makes **PetaDroid** models robust to code transformation and
obfuscation in general. We argue that training machine learning models on dynamic
fragments enhances the resiliency of the models against code transformation.

8.2.5   Change Over Time Resiliency

An important feature in modern Android malware detection is the resiliency to change over time [1, 2, 4]. We study the resiliency of **PetaDroid** over the last 7 years (2013–2019). We randomly sample from AndroZoo repository a number of $10k$ Android apps ($5k$ malware and $5k$ benign apps) for each year (2013–2019). As result, we have $70k = 35k_{Mal} + 35k_{Ben}$. We build the CNN ensemble using year Y_x samples and evaluate on the other years $Y_{1..N}$ samples. Figure 8.6a shows the general and the confidence performances of PetaDroid, for models trained on 2013 samples, in terms of F1-score on 2014–2019 samples. As shown in Fig. 8.6a, **PetaDroid** , trained on 2013 dataset, achieved $98.17, 96.10, 93.01, 70.60, 54.82$, and 55.59% F1-score on 2014, 2015, 2016, 2017, 2018, and 2019 datasets, respectively. **PetaDroid** sustains a relatively good performance over the first few years. In 2018 and 2019, the performance drops considerably. In comparison to MaMaDroid [2], **PetaDroid** shows a higher time resiliency over 7 years, while MaMaDroid drops considerably in year three (40% F1-score on year four). Figure 8.6b shows the training is on 2014 samples, which shows a performance enhancement over the overall evaluation period. The overall performance tends to increase as we train on a recent year dataset as depicted in Fig. 8.6c, d and e. In Figs. 8.6f, g, training is on samples from 2018 and 2019 respectively. **PetaDroid** performance slightly decreases on old samples from 2013 and 2014. Our interpretation is that old and deprecated Android APIs are not present in new apps from 2018 and 2019, which we use for the training, and this influences negatively the detection performance.

We take from this experiment that **PetaDroid** is resilient to change over time for years $t \pm 2$ when we train on year Y_t samples. **PetaDroid** covers about 5 years $\{Y_{t-2}, Y_{t-1}, Y_t, Y_{t+1}, Y_{t+2}\}$ of Android app change.

8.2.6   PetaDroid Automatic Adaptation

PetaDroid automatic adaptation goes beyond time resiliency. PetaDroid employs the confidence performance to collect an extension dataset X_{extend} during the deployment. **PetaDroid** automatically uses X_{extend} in addition to the previous build dataset as a new build dataset $X_{build(t)} = X_{build(t-1)} \cup X_{extend}$ to build a new ensemble at every new epoch. Table 8.9 depicts **PetaDroid** performance with and without automatic adaptation. **PetaDroid** achieves very good results compared to the previous section (Fig. 8.6a). PetaDroid maintains an F1-score in the range of 83–95% during all years. Without adaption, **PetaDroid** F1-score drops considerably starting from 2017 samples. Table 8.9 shows the performance of revisiting detection decisions on previous Android apps X_{test} (benign and malware) after updating **PetaDroid** ensemble using $X_{build} \cup X_{extend}, X_{extend} \subseteq X_{test}$. The update performance is significantly enhanced in the overall detection during all years. Revisiting malware detection decisions is common practice in App market

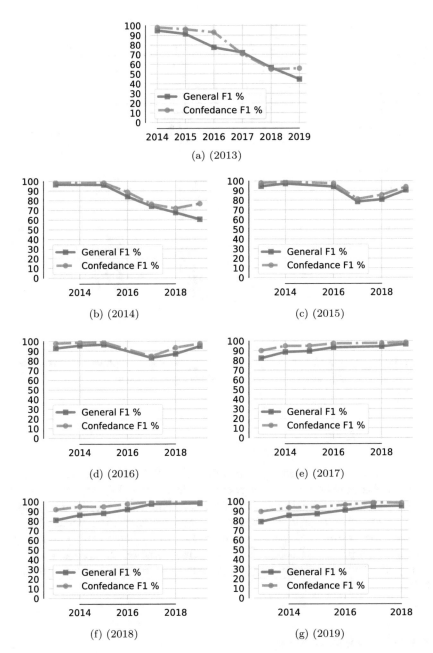

Fig. 8.6 PetaDroid resiliency to changes over time. (**a**) (2013). (**b**) (2014). (**c**) (2015). (**d**) (2016). (**e**) (2017). (**f**) (2018). (**g**) (2019)

Table 8.9 Performance of PetaDroid automatic adaptation

Year	No update (F1%)	General (F1%)	Confidence (F1%)	Update (F1%)
2014	98.2	97.0	97.9	99.7
2015	96.1	95.8	96.7	97.5
2016	93.0	93.3	94.8	96.4
2017	70.6	83.9	84.2	95.4
2018	54.8	87.6	91.6	93.8
2019	55.6	96.3	98.7	99.1

Fig. 8.7 PetaDroid runtime efficiency

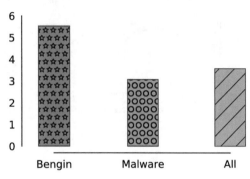

(periodic full or partial scan of the market's apps), which empowers the use case of PetaDroid automatic adaptation feature and the update metric.

8.2.7 *Efficiency*

In Fig. 8.7, we depict the average time of **PetaDroid** detection process. The latter includes disassembly, preprocessing, and inference time. **PetaDroid** spends, on average, 4.0 s to fingerprint an Android app. The runtime increases for benign apps, 5.5 s, because their package sizes tend to be larger compared to malicious ones. For malware apps, **PetaDroid** spends, on average, 3.0 s for fingerprinting on the app. The detection process of benign apps takes more time on the average compared with malicious apps because benign apps tend to have larger size than malicious apps.

8.3 Comparative Study

In this section, we conduct a comparative study between **PetaDroid** and state-of-the-art Android malware detection systems namely, **MaMaDroid** [1, 2] and **DroidAPIMiner** [19]. Our comparison is based on applying **PetaDroid** on the same dataset (malicious and benign apps) and using the same settings that **MaMaDroid** used in the evaluation (provided by the authors in [2]). The dataset is composed of 8.5K benign and 35.5K malicious apps in addition to Drebin [3] dataset. The malicious samples are tagged by time; malicious apps from 2012 (Drebin), 2013, 2014, 2015, and 2016 and benign apps are tagged as oldbenign and newbenign, according to MaMaDroid evaluation.

8.3.1 Detection Performance Comparison

Table 8.10 depicts the direct comparison between MaMaDroid and PetaDroid different dataset combinations. In PetaDroid, we present the general and the confidence performance in terms of F1-score. For MaMaDroid and DroidAPIMiner, we present the original evaluation result [2] in terms of F1 score, which are equivalent to the general performance in our case. Notice that, we present only the best results of MaMaDroid and DroidAPIMiner as reported in [2].

As depicted in Table 8.10, PetaDroid outperforms MaMaDroid on all datasets in the general performance. The detection performance gap increases with the confidence-based performance. Notice that the coverage in the confidence-based settings is almost perfect (only few apps have been filtered due to the low confidence) for all the experiments in Table 8.10.

8.3.2 Efficiency Comparison

In Table 8.11, we report the required average time for MaMaDroid and PetaDroid to fingerprint one Android app. PetaDroid takes 03.58 ± 04.21 s on average for the whole process (DEX disassembly, assembly preprocessing, CNN ensemble inference). MaMaDroid, compared to PetaDroid, tends to be slower due to the heavy preprocessing. MaMaDroid preprocessing [1] is composed of the call graph extraction, sequence extraction, and Markov change modeling, which require 25.40 ± 63.00, 1.73 ± 3.2, 6.7 ± 3.8 s respectively for benign samples and 09.20 ± 14.00, 1.67 ± 3.1, 2.5 ± 3.2 s respectively for malicious samples. On average, PetaDroid ($3.58s$) is approximately eight times faster than MaMaDroid (≈ 23).

Table 8.10 Detection performance of MaMaDroid, PetaDroid, and DroidAPIMiner

	General confidence		
	Peta (F1%)	MaMa (F1%)	Miner (F1%)
drebin&oldbenign	98.94–99.40	96.00	32.00
2013&oldbenign	99.43–99.81	97.00	36.00
2014&oldbenign	98.94–99.47	95.00	62.00
2014&newbenign	99.54–99.83	99.00	92.00
2015&newbenign	97.98–98.95	95.00	77.00
2016&newbenign	97.44–98.60	92.00	36.00

Table 8.11 MaMaDroid and PetaDroid runtime

	PetaDroid (seconds)	MaMaDroid (seconds)
Malware	02.64 ± 03.94	09.20 ± 14.00 + 1.67 ± 3.1 + 2.5 ± 3.2
Benign	05.54 ± 05.12	25.40 ± 63.00 + 1.73 ± 3.2 + 6.7 ± 3.8
Average	03.58 ± 04.21	≈23 s

8.3.3 Time Resiliency Comparison

MaMaDroid evaluation emphasizes the importance of time resiliency for modern Android malware detection. Table 8.12 depicts the performance with different dataset settings, such as training using an old malware dataset and testing on a newer one. PetaDroid outperforms (or obtains a very similar result in few cases) MaMaDroid and DroidAPIMiner in all settings. Furthermore, the results show that PetaDroid is more robust to time resiliency compared to MaMaDroid [2].

8.4 Case Studies

In this section, we conduct mega-scale experiments on AndroZoo dataset (9.5 million Android apps). We argue that these experiments reflect real world deployments due to the dataset size, time distribution (2010–2019), and malware family diversity. We report PetaDroid detection performance using our automatic adaptation feature.

8.4.1 Scalable Detection

In this experiment, we employ 8.5 out of 9.5 million Android apps from AndroZoo dataset. The used dataset is composed of 1.0 million malicious samples and 7.5 million benign sample. We filter out app samples that do not correlate with VirusTotal, or they have less than five maliciousness flags in VirusTotal. In our experiments, we randomly sample k samples as build dataset X_{build} and use the rest $8.5M - k$ as X_{test}. We use different k sizes, $k \in \{10k, 20k, 50k, 70k, 100k\}$, and we repeat each experiment ten times to compute the average detection performance. In Table 8.13, we report the detection performance in terms of F1-score of PetaDroid on AndroZoo dataset. PetaDroid shows a high F1-score for all the experiments, $95 - 97\%$ F1-score. We achieved 95.34% F1-score when the build set is only $10k$. We argue that fragment randomization plays an important role in achieving these detection results because it acts as a data augmenter (the randomization generates several canonical instruction sequences from a given Android app through the permutation of the methods code) during the training phase.

Table 8.12 Classification performance of MaMaDroid, PetaDroid, DroidAPIMiner

Testing sets	drebin & oldbenign			2013 & oldbenign			2014 & oldbenign			2015 & oldbenign			2016 & oldbenign		
Training sets	Miner	MaMa	Peta	Miner	MaMa	Peta	Miner	MaMa	Peta	Miner	MaMa	Peta	Miner	MaMa	Peta
drebin&oldbenign	32.0	96.0	**99.4**	35.0	95.0	**98.6**	34.0	72.0	**77.5**	30.0	39.0	**44.0**	33.0	42.0	**47.0**
2013&oldbenign	33.0	94.0	**97.8**	36.0	97.0	**99.6**	35.0	73.0	**85.4**	31.0	37.0	**59.3**	33.0	28.0	**56.6**
2014&oldbenign	36.0	92.0	**95.8**	39.0	93.0	**98.6**	62.0	95.0	**99.4**	33.0	78.0	**91.4**	37.0	75.0	**88.9**

Training sets	drebin & newbenign			2013 & newbenign			2014 & newbenign			2015 & newbenign			2016 & newbenign		
	Miner	MaMa	Peta	Miner	MaMa	Peta	Miner	MaMa	Peta	Miner	MaMa	Peta	Miner	MaMa	Peta
2014&newbenign	76.0	98.0	**99.3**	75.0	98.0	**99.7**	92.0	99.0	**99.8**	67.0	85.0	**91.4**	65.0	81.0	**82.1**
2015&newbenign	68.0	97.0	**97.1**	68.0	97.0	**97.8**	69.0	**99.0**	98.9	77.0	95.0	**99.0**	65.0	88.0	**95.4**
2016&newbenign	33.0	**96.0**	95.6	35.0	98.0	**98.2**	36.0	**98.0**	97.9	34.0	92.0	**95.2**	36.0	92.0	**98.3**

The bold values represent the values of PetaDroid that are the best compared to other techniques.

Table 8.13 PetaDroid mega-scale detection performance

#samples	General (F1 %)	Confidence (F1 %)
10k	95.34	97.88
20k	96.17	98.01
50k	96.50	98.10
70k	96.76	98.11
100k	97.04	98.17

Table 8.14 Autonomous adaptation on mega-scale dataset

Update epoch	Before update (F1 %)		After update (F1 %)	
	General	Confidence	General	Confidence
2013-01-31	/	/	/	/
2013-04-30	96.02	98.43	99.01	99.71
2013-07-31	94.52	96.12	97.95	99.56
2013-10-31	94.42	97.37	97.03	99.56
2014-01-31	83.45	92.74	95.45	99.37
2014-04-30	90.48	96.21	94.15	99.43
2014-07-31	86.98	95.79	91.53	99.11
2014-10-31	92.32	98.47	93.11	99.00
2015-01-31	91.57	97.72	90.91	99.18
2015-04-30	91.31	98.55	92.72	99.09
2015-07-31	88.16	97.46	88.90	98.99
2015-10-31	73.82	87.45	83.44	97.57
2016-01-31	78.59	90.92	85.11	96.26
2016-04-30	84.78	95.44	86.38	98.33
2016-07-31	71.39	88.08	80.27	93.54
2016-10-31	78.31	85.79	79.68	90.75

8.4.2   Scalable Automatic Adaptation

In this experiment, we put the automatic adaptation feature on mega-scale test using 5.5 million samples from AndroZoo dataset (2013–2016) on 25 training epochs (every 3 months). We initiate **PetaDroid** on only $25k$ build datasets collected between 2013-Jan-01 and 2013-Jan-31. **PetaDroid** rebuilds new CNN ensemble for each 3 month samples by retraining on $X_{build(t)} = X_{build(t-1)} \cap X_{extend}$.

In Table 8.14, we report the general and confidence performance before and after updating **PetaDroid** CNN ensemble on an extended build dataset. The automatic adaption feature achieves very good results. The general and confidence-based performance in terms of F1 score vary between 71.39–96.02% and 85.79–98.55%, respectively. These performance results increase considerably (90.75–99.71% F1-score) after revising the previous detection decisions using an updated CNN ensemble using a new X_{extend} on each epoch.

8.5 Summary

In this chapter, we presented PetaDroid, an Android malware detection, and family clustering framework for large-scale deployments. PetaDroid employs supervised machine learning, an ensemble of convolutional neural networks on top of *Inst2Vec* features, to fingerprint Android malicious apps accurately. Furthermore, PetaDroid uses unsupervised machine learning, precisely DBScan clustering on top of *InstNGram2Vec* and deep auto-encoders features, to cluster highly similar malicious apps into their most likely malware family groups. In PetaDroid, we introduced fragment-based detection, in which we randomize the macro-action of Android APIs while keeping the inner order of methods' sequences. Fragment randomization acts as a data augmentation mechanism during the training and strengthens detection robustness against common obfuscation techniques during deployment. Also, we introduced the automatic adaption technique that leverages confidence-based decision making to build a new CNN ensemble on confidence detection samples. The adaptation technique automatically enhances PetaDroid time resiliency. We conducted a thorough evaluation of different reference datasets and various settings. PetaDroid achieved high detection (98–99% F1-score) and family clustering (96% cluster homogeneity) performance. Our comparative study between PetaDroid and MaMaDroid [1, 2] shows that PetaDroid outperforms state-of-the-art solutions on various settings. We evaluate PetaDroid on a market scale Android dataset, over $100TB$ of data and 9.7 million samples.

References

1. E. Mariconti, L. Onwuzurike, P. Andriotis, E.D. Cristofaro, G.J. Ross, G. Stringhini, Mamadroid: detecting android malware by building Markov chains of behavioral models, in *24th Annual Network and Distributed System Security Symposium, NDSS 2017, San Diego, California, USA, February 26 - March 1, 2017* (2017)
2. L. Onwuzurike, E. Mariconti, P. Andriotis, E.D. Cristofaro, G.J. Ross, G. Stringhini, Mamadroid: detecting android malware by building Markov chains of behavioral models (extended version). ACM Trans. Priv. Secur. **22**(2), 14:1–14:34 (2019)
3. D. Arp, M. Spreitzenbarth, M. Hubner, H. Gascon, K. Rieck, DREBIN: effective and explainable detection of android malware in your pocket, in *21st Annual Network and Distributed System Security Symposium, NDSS 2014, San Diego, California, USA, February 23–26, 2014* (2014)
4. E.B. Karbab, M. Debbabi, A. Derhab, D. Mouheb, Maldozer: automatic framework for android malware detection using deep learning. Digit. Invest. **24**, S48–S59 (2018)
5. G. Suarez-Tangil, S.K. Dash, M. Ahmadi, J. Kinder, G. Giacinto, L. Cavallaro, DroidSieve: fast and accurate classification of obfuscated android malware (2017), pp. 309–320
6. S. Chen, M. Xue, Z. Tang, L. Xu, H. Zhu, Stormdroid: a streaminglized machine learning-based system for detecting android malware, in *Proceedings of the 11th ACM on Asia Conference on Computer and Communications Security, AsiaCCS 2016, Xi'an, China, May 30 - June 3, 2016* (2016), pp. 377–388

7. T. Mikolov, I. Sutskever, K. Chen, G.S. Corrado, J. Dean, Distributed representations of words and phrases and their compositionality, in *Advances in Neural Information Processing Systems 26: 27th Annual Conference on Neural Information Processing Systems 2013. Proceedings of a Meeting Held December 5–8, 2013, Lake Tahoe, Nevada, United States* (2013), pp. 3111–3119

8. G.E. Hinton, A. Krizhevsky, S.D. Wang, Transforming auto-encoders, in *Artificial Neural Networks and Machine Learning - ICANN 2011 - 21st International Conference on Artificial Neural Networks, Espoo, Finland, June 14–17, 2011, Proceedings, Part I* (2011), pp. 44–51

9. N. McLaughlin, J.M. del Rincón, B. Kang, S.Y. Yerima, P.C. Miller, S. Sezer, Y. Safaei, E. Trickel, Z. Zhao, A. Doupé, G. Ahn, Deep android malware detection, in *Proceedings of the Seventh ACM on Conference on Data and Application Security and Privacy, CODASPY 2017, Scottsdale, AZ, USA, March 22–24, 2017* (2017), pp. 301–308

10. S.H.H. Ding, B.C.M. Fung, P. Charland, Asm2vec: boosting static representation robustness for binary clone search against code obfuscation and compiler optimization, in *2019 IEEE Symposium on Security and Privacy, SP 2019, San Francisco, CA, USA, May 19–23, 2019* (2019), pp. 472–489

11. Y. Kim, Convolutional neural networks for sentence classification, in *Proceedings of the 2014 Conference on Empirical Methods in Natural Language Processing, EMNLP 2014, October 25–29, 2014, Doha, Qatar, A Meeting of SIGDAT, a Special Interest Group of the ACL* (2014), pp. 1746–1751

12. X. Zhang, J.J. Zhao, Y. LeCun, Character-level convolutional networks for text classification, in *Advances in Neural Information Processing Systems 28: Annual Conference on Neural Information Processing Systems 2015, December 7–12, 2015, Montreal, Quebec, Canada* (2015), pp. 649–657

13. I. Goodfellow, Y. Bengio, A. Courville, *Deep Learning* (MIT Press, Cambridge, 2016). http://www.deeplearningbook.org

14. T. Abou-Assaleh, N. Cercone, V. Keselj, R. Sweidan, N-gram-based detection of new malicious code, in *28th International Computer Software and Applications Conference (COMPSAC 2004), Design and Assessment of Trustworthy Software-Based Systems, 27–30 September 2004, Hong Kong, China, Workshop Papers* (2004), pp. 41–42

15. Q. Shi, J. Petterson, G. Dror, J. Langford, A.J. Smola, S.V.N. Vishwanathan, Hash kernels for structured data. J. Mach. Learn. Res. **10**, 2615–2637 (2009)

16. Y. Zhou, X. Jiang, Dissecting android malware: characterization and evolution, in *IEEE Symposium on Security and Privacy, SP 2012, 21–23 May 2012, San Francisco, California, USA* (2012), pp. 95–109

17. F. Wei, Y. Li, S. Roy, X. Ou, W. Zhou, Deep ground truth analysis of current android malware, in *Detection of Intrusions and Malware, and Vulnerability Assessment - 14th International Conference, DIMVA 2017, Bonn, Germany, July 6–7, 2017, Proceedings* (2017), pp. 252–276

18. K. Allix, T.F. Bissyandé, J. Klein, Y.L. Traon, Androzoo: collecting millions of android apps for the research community, in *Proceedings of the 13th International Conference on Mining Software Repositories, MSR 2016, Austin, TX, USA, May 14–22, 2016* (2016), pp. 468–471

19. Y. Aafer, W. Du, H. Yin, Droidapiminer: mining API-level features for robust malware detection in android, in *Security and Privacy in Communication Networks - 9th International ICST Conference, SecureComm 2013, Sydney, NSW, Australia, September 25–28, 2013, Revised Selected Papers* (2013), pp. 86–103

20. D. Maiorca, D. Ariu, I. Corona, M. Aresu, G. Giacinto, Stealth attacks: an extended insight into the obfuscation effects on android malware. Comput. Secur. **51**, 16–31 (2015)

21. V. Rastogi, Y. Chen, X. Jiang, DroidChameleon: evaluating android anti-malware against transformation attacks, in *8th ACM Symposium on Information, Computer and Communications Security, ASIA CCS'13, Hangzhou, China - May 08 - 10, 2013* (2013), pp. 329–334

22. A. Rosenberg, J. Hirschberg, V-measure: a conditional entropy-based external cluster evaluation measure, in *EMNLP-CoNLL 2007, Proceedings of the 2007 Joint Conference on Empirical Methods in Natural Language Processing and Computational Natural Language Learning, June 28–30, 2007, Prague, Czech Republic* (2007), pp. 410–420

Chapter 9
Conclusion

9.1 Concluding Remarks

At the heart of the rapid growth of software technologies, the development of mobile apps enhances both economic and social interactions. Mobile apps running on smart devices are nowadays ubiquitous due to their convenience. For instance, users can presently use apps as Google Pay service to purchase products online and to make payments in retail stores. However, the growth of the mobile market apps has increased the concerns about the security of the apps. Android [1] is widely adopted mobile OS in smart devices, especially in the emerging Internet of Things (IoT) world through Android Thing [2], an Android-based IoT system. Unfortunately, significant amounts of malicious software or malware, which are developed for a variety of purposes, aim at disrupting the well being of existing systems across many software platforms and hardware architectures. For example, 1,548,129 and 2,333,777 new Android malware were discovered [3] in 2014 and 2015, respectively. Nowadays, the number of malware samples reaches millions per month and is growing exponentially over time. In this context, it is a desideratum to elaborate scalable, robust, and accurate techniques and frameworks that tackle two specific problems: (1) Malware detection—distinguishing malicious from benign applications, and (2) malware family attribution—assigning malware samples to known families.

This book is dedicated to tackling Android malware fingerprinting, detection, and family attribution, by proposing a series of frameworks and techniques to detect and attribute Android malware samples. Android malware detection was the main objective of our elaborated frameworks and systems. However, the core techniques and methods employed by these systems have potential application to general malware fingerprinting. The elaborated frameworks demonstrated very competitive Android malware fingerprinting results surpassing state-of-the-art solutions available at the time of writing this book. More specifically, the book presented the following solutions:

© The Author(s), under exclusive license to Springer Nature Switzerland AG 2021   195
E. B. Karbab et al., *Android Malware Detection Using Machine Learning*, Advances in Information Security 86, https://doi.org/10.1007/978-3-030-74664-3_9

- **APK-DNA** (Chap. 4): We elaborated a versatile approximate fingerprint [4] to capture the maximum information from the static content of an Android malware sample.
- **Cypider** (Chap. 4): We proposed a scalable malware clustering technique [5] leveraging **APK-DNA** approximate fingerprints along with graph partitioning techniques.
- **MalDy** (Chap. 5): We elaborated a supervised machine learning approach [6, 7] for malware detection on top of **DySign** fingerprints using dynamic analysis features.
- **ToGather** (Chap. 6): We proposed a cyber-infrastructure detector [8, 9] for Android malware in the cyber-space starting from network information of Android malware as well as static and dynamic analyses.
- **MalDozer** (Chap. 7): We proposed a portable and automatic Android malware detection and family attribution framework [10–12] that relies on sequences classification using deep learning techniques.
- **PetaDroid** (Chap. 8): We proposed a novel framework for Android malware detection that enhances the resiliency to code transformation and common obfuscation methods by input randomization. Also, we leveraged confidence-based detection to build new machine learning detection models that are able to adapt to new benign and malicious apps.

9.2   Lessons Learned

We summarize in the following the main lessons that have been learned in this book:

- *Representation learning enables scalability*: Representation learning is at the core of malware automatic feature engineering. In our context, representation learning is an automatic and a data-driven process for generating malware embeddings. For example, the use of the word embedding (word2vec) technique helps learning the underlying semantics in an unsupervised manner. Existing large malware corpuses and unsupervised representation learning techniques help generating precise malware embeddings. More importantly, the precision of embeddings increases with the size of the malware corpus.
- *Natural language processing is a key*: We learned the usefulness of leveraging NLP abstractions in malware code analysis. Virtually, most NLP techniques used to segregate and analyze natural language are usable in the context of malware fingerprinting and detection. We believe that NLP abstractions and techniques are key for modern malware detection.
- *Machine learning is crucial*: Machine learning techniques are essential in elaborating advanced malware detection solutions. All existing state-of-the-art malware detection solutions rely nowadays on machine learning as a workhorse to fingerprint malware. Deep learning techniques have the edge over classical

ones due to automatic discovery and filtering of relevant malware features from raw malware content.

- *Obfuscation learning is possible*: We discovered the possibility to automatically learn obfuscation pattern through supervised techniques. Providing obfuscated malware samples to the training process allows learning common obfuscation methods while improving the overall generalization of the produced models.

9.3   Future Research Directions

In the following, we discuss potential future research directions:

- *Obfuscation on other platforms*: We have evaluated the proposed Android malware detection frameworks on different obfuscation techniques. Our frameworks show high detection performance on obfuscated Android malware. However, there is a need to evaluate our obfuscation resilient techniques on other platforms' obfuscated samples.
- *Tackling advanced obfuscations techniques*: In the context of this book, we have carried out our evaluations on common obfuscation and code transformation techniques. As a future research direction, we could investigate the robustness of the proposed frameworks and techniques on advanced obfuscation techniques that employ heavy code transformation such as *control flow flattening*.
- *Additional deep learning techniques*: We have employed different machine/deep learning techniques to fingerprint and detect Android malware. Exploring additional deep learning techniques can provide an important future direction for malware detection in general.
- *Network traffic features*: Throughout this book, we have mainly employed dynamic and static analyses features to fingerprint Android malware. As a future research direction, we aim at engaging network traffic inspection as another source of features for Android malware detection.

References

1. Android Operating System, https://www.android.com/. Accessed Jan 2019
2. Android Things, https://developer.android.com/things/. Accessed Sept 2016
3. G DATA Mobile Malware Report, https://public.gdatasoftware.com/Presse/Publikationen/ Malware_Reports/US/G_DATA_MobileMWR_Q4_2015_US.pdf. Accessed Dec 2016
4. E.B. Karbab, M. Debbabi, D. Mouheb, Fingerprinting Android packaging: generating DNAs for malware detection. Digit. Investig. **18**, S33–S45 (2016)
5. E.B. Karbab, M. Debbabi, A. Derhab, D. Mouheb, Cypider: building community-based cyber-defense infrastructure for android malware detection, in *Proceedings of the 32nd Annual Conference on Computer Security Applications, ACSAC 2016, Los Angeles, CA, USA*, 5–9 Dec 2016, pp. 348–362

6. E.B. Karbab, M. Debbabi, MalDy: portable, data-driven malware detection using natural language processing and machine learning techniques on behavioral analysis reports. Digit. Investig. **28**, S77–S87 (2019)

7. E.B. Karbab, M. Debbabi, Portable, data-driven malware detection using language processing and machine learning techniques on behavioral analysis reports. CoRR, abs/1812.10327 (2018)

8. E.B. Karbab, M. Debbabi, ToGather: automatic investigation of android malware cyber-infrastructures, in *Proceedings of the 13th International Conference on Availability, Reliability and Security, ARES 2018, Hamburg, Germany*, 27–30 Aug 2018, pp. 20:1–20:10

9. E.B. Karbab, M. Debbabi, Automatic investigation framework for android malware cyber-infrastructures. CoRR, abs/1806.08893 (2018)

10. E.B. Karbab, M. Debbabi, A. Derhab, D. Mouheb, Android malware detection using deep learning on API method sequences. CoRR, abs/1712.08996 (2017)

11. S. Alrabaee, E.B. Karbab, L. Wang, M. Debbabi, BinEye: towards efficient binary authorship characterization using deep learning, in *Computer Security - ESORICS 2019 - Proceedings, Part II 24th European Symposium on Research in Computer Security, Luxembourg*, 23–27 Sept 2019, pp. 47–67

12. E.B. Karbab, M. Debbabi, A. Derhab, D. Mouheb, MalDozer: automatic framework for android malware detection using deep learning. Digit. Investig. **24**, S48–S59 (2018)

Index

199

Printed in the United States
by Baker & Taylor Publisher Services